SHAPING CANADA'S ENVIRONMENT

General Editor—James Forrester
Supervisor of Geography
The Board of Education for the City of Hamilton

SHAPING CANADA'S ENVIRONMENT

Building for People

Freeway and Downtown—
New Frameworks for Modern Needs

RONALD E. RICHARDSON
Editor, *Heavy Construction News*
GEORGE H. McNEVIN
Technical Field Editor, *Heavy Construction News*
WALTER G. ROOKE
Associate Editor, *Heavy Construction News*

THE RYERSON PRESS   MACLEAN-HUNTER LIMITED
Toronto, Winnipeg, Vancouver   Toronto

© The Ryerson Press and Maclean-Hunter, Ltd., 1970

ISBN 0-7700-3207-9

Acknowledgments

The publishers wish to acknowledge with gratitude those who have given their permission to use copyrighted material in this book. Every effort has been made to credit all sources correctly. The authors and publishers will welcome any information that will allow them to correct any errors or omissions.

HIGHWAY FOR TODAY: ONTARIO'S 401
The authors wish to acknowledge the cooperation of the staff of the Department of Highways, Ontario, in researching and providing materials; with particular thanks to the contribution of DHO staff writer Robert Baigent. Appreciation also goes to Zillah Ferguson for her efforts in manuscript preparation.

A CITY AND ITS HEART
The assistance provided by the managements of Trizec Corp., Place Bonaventure Inc., Place Victoria-St. Jacques Co. Inc., Canadian National Railways, CP Rail, and the Montreal City Planning and Housing departments is gratefully acknowledged. Particularly helpful in numerous interviews were Vincent Ponte, private city planning consultant; James Soden, president, Trizec Corp.; Thomas Phalen, president, Concordia Construction Inc.; W. G. Kent, then vice-president, leasing, Place Victoria-St. Jacques Co. Inc.; J. E. Brett, Montreal, structural engineer on the Place Ville Marie project; Anshell Melamed, Montreal City Planning Department and Donovan Pinker, city planning consultant. Helpful research was provided by Toronto City Development Commissioner D. Graham Emsley, C. A. Hughes of his department, and Vithal Rahjan, Montreal. Special thanks are due to the two Christines, Joan, Bella and Betty (who work in the super blocks) for an afternoon of interviewing and their employers who made their time available.

INDUSTRY ON THE MOVE
The authors wish to thank the numerous manufacturing companies, engineering firms and government agencies who aided in compilation of the facts and projections expressed in this chapter. Special mention must be made of the Westinghouse Air Brake Co., Koehring Manufacturing Co., CP Rail and the General Motors of Canada, Ltd. Many technical equipment details and much historical information were provided by Rolland L. Jerry.

Cover: Walter G. Rooke
Title page and facing page 1: Department of Highways, Ontario.
Facing page 53: Walter G. Rooke.

PRINTED AND BOUND IN CANADA
BY THE RYERSON PRESS, TORONTO

Foreword

Our concept of resources has until recently dealt largely with commodities, with tangible things. Too often we have overlooked the human and cultural resources which are all around us. Yet human effort and ingenuity, and the ability to use them well, are the greatest resources man possesses.

The central theme of this book is the story of modern technology and how it is reshaping our way of life. The construction industry in Canada has profoundly affected the lives of people. It is both a determinant and an indicator of the direction of the economy. It touches business, commerce, transportation, manufacturing, worship, home, recreation and entertainment. It changes behavioural patterns.

The chapter on Ontario's Highway 401 looks at the effect of the highway upon economic, political and social life. By the end of the Second World War, Ontario had grown far beyond the stage where local transportation was the main function of the provincial highways. More and more concern was being shown for the fast, efficient, long distance movement of people and goods. The increasing number and speed of automobiles and trucks necessitated a new concept in highway construction. The builders of Highway 401 tried to meet these requirements by meticulous research and careful planning.

The designers and planners were able to arrange the road pattern to give almost the shortest distance between the points to be connected. The construction men had the technical knowledge and the mechanical equipment to cross rivers easily, fill swamps if they so desired and to ignore other topographic features where they deemed it necessary.

But what of the effects of Highway 401 upon the way of life of the people of Southern Ontario? The planners knew that transportation creates place utility. Did they expect the building of Highway 401 to underline this principle? Did they imagine that their super-highway would give hundreds of thousands of people a comparatively free choice of residence, place of work and shopping facilities, all possibly many miles apart? Could they have forecasted the effect this new high speed traffic corridor across Southern Ontario would have on industrial location, land use patterns and distance from home to work? Did they foresee that some places along the new route would stagnate while most would flourish?

In the Montreal chapter the aim has been to examine the metamorph- osis of the downtown core. Montreal may well be Canada's most exciting city and in some ways it is a world leader in changing urban form and urban living. In the area of separation of forms of movement, the city has no peer. Montreal's system provides separate surfaces for all forms of movement. In its "streets" and malls, the pedestrian is permitted complete forgetfulness of foul weather and traffic hazards. Its climate- controlled "street" system is made up not only of subterranean tunnels but also of passages following original grade levels. Natural daylight enters at intervals.

In the past, most cities were characterized by inflexible single use zoning patterns. Planners of today have noted that a multi-use building represents a desirable break from that trend. The idea of using one building for many different purposes has gained favour in Montreal. Today, Place Bonaventure incorporates retailing facilities, transportation services, office space, an international merchandising centre, a huge trade fair floor and a 400-room hotel, and it is only one of several multi-use complexes.

Vincent Ponte, the private town planning consultant who worked with the owners and architects in the design of both Place Ville Marie and Place Bonaventure, has suggested the introduction of municipal bylaws that would require the owners of new buildings in the central business district to provide pedestrian ways above or below ground that could be plugged into the pedestrian grid as it approaches. He notes that zoning in downtown areas today chiefly concerns itself with controlling land use, density and with providing sufficient light and air for streets. Thus zoning's effect is vertical while the multilevel city core is concerned with horizontals, of decks and levels extending blockwise across the core, over and under existing streets. In a country of climatic extremes such as Canada, surely such ideas as have been put into practice in Montreal should spread to other cities.

The final chapter traces the development of the construction industry in Canada, discusses the value of the industry to the Canadian economy, and outlines the industry's structure.

J.F., April 1970.

Contents

CHAPTER 1

Highway for Today: Ontario's 401

Highway 401 is the most important single development changing the economic and social patterns of Ontario.

Professor E. G. Pleva,
University of Western Ontario,
March, 1963.

". . . And now, let's go up to our traffic helicopter and check on road conditions this morning in and around the city."

The radio announcer's voice reaches out to thousands of listeners, but the most attentive at this moment are the car commuters. Some have been on the road for almost an hour, others just a few minutes, but all are concerned with the flow of vehicles along the routes taking them from home to work.

The automobile and the highway are an integral part of their lives. They rely on them for getting to work on time, often from distances that would have been considered impractical, if not impossible, only a few years before.

Today's network of super-highways has transformed social and economic patterns. And as these motor-age pilgrims leave their homes each morning they anticipate a smooth, trouble-free journey. Generally the trip does go smoothly, with a minimum of delays and slow-downs—testimony both to the efficiency of the machine age and the automobile manufacturers and to the skill of the highway planners.

As the control room operator at the radio station flips switches, there is a crackle of static, then a voice sounds clearly over the muted roar of the helicopter engine.

"At the moment we are flying west along the Macdonald-Cartier Freeway, just over the Yonge Street interchange. Everything looks fine from up here, with traffic moving along at a steady pace and only a slight slowdown at the exit ramps. We'll be reporting back in a few minutes. Now back to our studios."

The size and scope of the Macdonald-Cartier Freeway (Highway 401) are impressive both from the air and at ground level as it cuts across the northern part of Toronto, Canada's second largest metropolis. Wide lanes of pavement, spanned by soaring, split-level interchanges, provide the setting for the flow of vehicles. There is an around-the-clock hum of traffic as over 100,000 vehicles a day stream along the bypass, but it is during the peak rush-hour periods of morning and late afternoon that the steel and concrete corridors are most alive with cars and trucks. And every driver, although backed-up by the best built-in safety features highway designers can produce, must still muster the skills to manoeuvre his vehicle in and out of the whirl of traffic at speeds up to 70 miles an hour.

The very success of such roads as the 401 is now creating problems—both mechanical and environmental—from greatly increased volumes of traffic. It may be that combinations of mass transit and expressways will be the answer to the problem of moving people efficiently in urban areas; possibilities are discussed on pages 46 to 51.

TRANS-PROVINCIAL MOTORWAY: THE NEED

The story of the freeway extends far beyond the heavily-travelled Toronto bypass. The route sweeps 510 miles across the southern section of Ontario, from Windsor and the United States border city of Detroit in the west, to the Quebec border and Montreal-bound Route 20 in the east.

The Macdonald-Cartier Freeway (so-named prior to centennial celebrations to honour two of the Fathers of Confederation, Sir John A. Macdonald and Sir Georges-Etienne Cartier) serves not only as the surface for the quick mobility of people, but is equally as important as an expediter of goods. It has an almost magnetic attraction for industry.

Yet the designers deliberately set out to bypass communities with a "controlled-access" route to avoid the congestion created by through traffic tangling with local traffic. This meant eliminating all level crossings, private driveways and intersections. Drivers would enter or leave the highways only at specified interchanges. All roads that had to cross the path of the main highway would be carried either over or under the right-of-way.

This design concept produced a freeway, not only in the sense that it was toll free (some major highways are financed on a charge per mile basis) but in terms of being free from trouble spots.

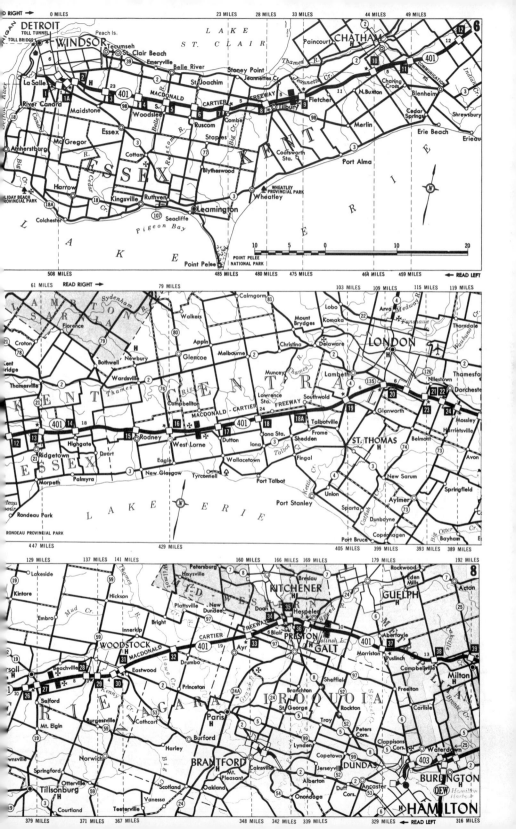

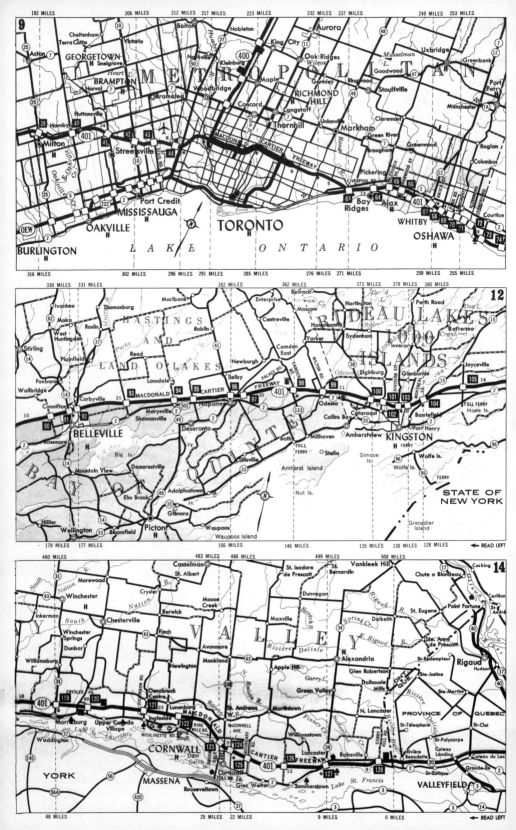

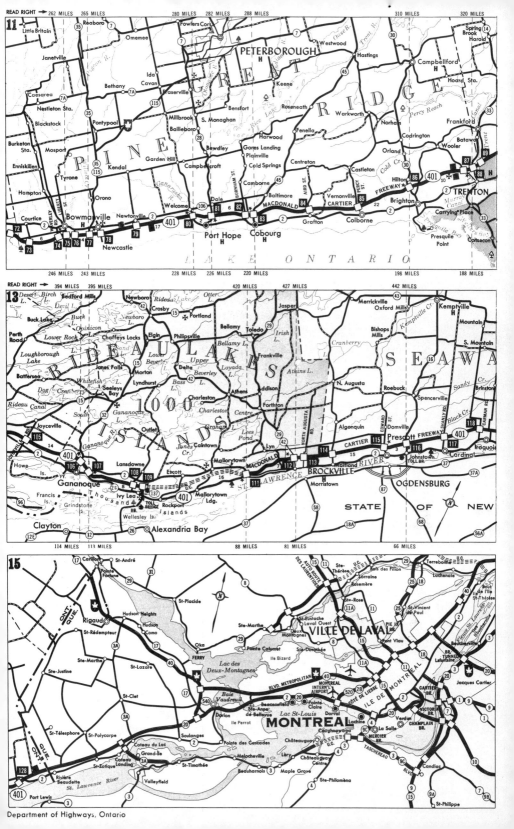

Designers followed these standards with the building of the 401, although the total concept was not possible at the outset. The control of access is extended each year by the building of more structures to separate intersecting roads and by the construction of new interchanges to connect additional main roads with the freeway. And with traffic volumes rising steadily, the basic four-lane, divided route will some day be expanded along its full length, as has been the case with the now 12-lane bypass in Metro Toronto.

Several years before the 401 was planned as a major, controlled-access route by the Department of Highways, Ontario (DHO), the department had already set a precedent with the Queen Elizabeth Way (QEW).

The QEW was designed to run 96 miles from Toronto, across the Niagara Peninsula to Fort Erie and, via the Peace Bridge, to connect with Buffalo, New York. Although the highway was not completed until after the Second World War, the first section from Toronto to Niagara Falls was officially opened in June, 1939.

At that time it was a shining example of what a modern highway should be. Incorporating many advances in road design, it ranked near the top among North American super-highways. The four-lane route was almost certainly the first of its kind in Canada and its cloverleaf at Highway 10, Mississauga, was the forerunner of all such interchanges in the country.

Guide for future planning

Before the first section of the QEW had been completed, DHO engineers realized that something more would be necessary to achieve the level of service and safety they were seeking. The great stumbling block with the QEW was the discovery that not even the provincial government had sufficient control of access to already-established properties abutting an existing road. The only answer was to construct parallel service roads on either side of the controlled-access highway—at the DHO's expense. A good deal of the QEW, particularly from Toronto to Hamilton, was constructed on a new right-of-way, but beyond Hamilton and into the Niagara region, sections of existing roads were incorporated.

From this experience, the department decided that in future it would create its own right-of-way, independent of any other roads in the area, for all controlled-access mileage.

Department of Highways, Ontario

This cloverleaf on the Queen Elizabeth Way at Highway 10 in Mississauga, just west of Metro Toronto, was the forerunner of all such Canadian interchanges. Opened in June, 1939, it combined design features for safe exits and entrances still widely used today. This photograph was taken about 1953.

By far the most important of the new routes to be designed within this concept has been Highway 401. Back in those pre-war years it was a vision of staggering dimensions; nothing less than a trans-provincial freeway, nearly six times the length of the QEW.

In the late 1930's, the main route from Windsor to the Quebec border was Highway 2. It carried more traffic in Ontario than any other King's Highway. (King's Highway is the designation of most main routes in the province.) And because it was the old "main" road in the province, it cut through cities, towns and villages, with all sorts of business enterprises and private dwellings adjacent to its right-of-way. As a result, there was serious traffic congestion along many parts of its 540-mile distance.

Applying the experience gained from the QEW, there was no attempt to convert this existing highway into a controlled-access route.

Department of Highways, Ontario

Before construction of the 401, most long distance traffic had to travel through city business sections. This view of Highway 2 through Trenton during the mid-1950's is typical of the traffic congestion which resulted. Tie-ups were particularly severe on week-ends when through traffic tangled with Saturday shoppers.

There would have to be a new highway, superior both in design and function to the QEW. Precisely where to locate such mileage was another question, for a four-lane, divided, controlled-access construction can cost up to 10 times as much as an ordinary two-lane route.

That was the size of the order facing the DHO planners in 1938. The stage was set. Planning was under way and some property had been purchased for the right-of-way. Then came the Second World War, and all thoughts of construction were put aside for the time being.

The war did, however, produce one beneficial side-effect. It allowed longer observation of the first controlled-access construction, the QEW. DHO planners learned, for instance, that the 132-foot right-of-way was not wide enough, and road design experts fixed 300 feet as a standard specification for all future highways in this category.

They also decided there would be no trees or shrubs in the median (the central area separating the traffic lanes). Although shrubs provided a pleasant view and filtered the glare of headlights, they often proved hazardous if vehicles went out of control. They also cut down the amount of snow storage space for snow ploughs and, in summer, increased the cost of mowing the grass. The planners decided that on new four-lane highways, they would restrict plants to the area between the shoulders and the outside limits of the right-of-way. They would also depress the grassed medians, for safety as well as for drainage.

In selecting the approximate corridor for the new highway, the main priority was to bypass as many communities as possible, in order to provide a high-speed throughway. At the same time, since Highway 2 passed through all the largest cities in southern Ontario, from Windsor to the Quebec border, it was necessary to locate the new highway close to all or most of these centres. Another priority was to lift some of the load from other highways running close to the 401.

Finally, the planners had to take into account the pressure exerted on the King's Highway system throughout this trans-provincial corridor by four major areas of population, three outside the province and one within. These were Detroit in the west, Montreal in the east and Buffalo, with the Niagara Peninsula border crossings from the south, about midway between the two. In Ontario itself the immediate highway problem was how to bypass Toronto and its burgeoning industrial and residential suburbs.

As the Second World War continued, the DHO undertook planning studies and other preparatory work. Although there were to be later revisions, a general route was determined for the full distance.

With the 401, the concept of the super-highway and the development of the construction equipment industry came together at just the right time. Before the war, the limited size of construction machinery and the great reliance on manual labour would have made the 401 virtually impossible due to cost.

But, in the post-war 1940's, there were large earth scrapers, bulldozers and trucks, high-capacity air compressors and rock drills, and dozens of other equipment items which had been developed during the war years.

The larger capacity of these units meant that vital work on other highway projects in the province could continue without causing an undue strain on the machinery and manpower resources of the construction industry. Important as the 401 was, it could not be allowed to hold up other work.

Improved machinery also meant expenses could be kept down. The cost of moving a cubic yard of earth remained at about 40 cents from 1920 to 1960. Similarly, the cost of removing rock was almost the same in 1960 as it had been in 1940. On the other hand, the manual expenses had soared. In 1920, it cost about 90 cents to lay 100 bricks; in 1960 the figure was about $6.40—a 700% increase.

Overloaded highways

From the outset, departmental policy was to build sections where the need was greatest, instead of starting at either end and plugging away in a continuous line until the full route was complete.

The DHO decided to construct only self-sufficient sections that would perform the greatest measure of service from the time they were opened. This was achieved by building, wherever possible, sections linked at both ends with King's Highway. Although those in charge of scheduling construction were sympathetic to local needs and desires, they were coolly analytical in adhering to the priority ratings that had been painstakingly given to every section of the route. Each year's programme was based on the expectation of the sums to be voted to the DHO by the provincial legislature. All major highways in Ontario are constructed and maintained by the provincial government from tax revenues—a large percentage from a tax on each gallon of gasoline sold.

The priority set for sections was criticized, generally by those whose region had not been chosen for the next area of construction. Likewise, there were those who questioned the precise route chosen. Sometimes it was a case of wanting the highway conveniently closer. Often it was the opposite; people were upset because the highway would slice off sections of the old homestead, or cut through a region they wanted to remain unchanged.

Traffic congestion

In December, 1947, the first four-lane, controlled-access section was placed in service. Two 22-foot wide strips of concrete pavement connected 18½ miles between Highland Creek in the east end of what is now Metro Toronto and Ritson Road in Oshawa. It was only a short distance up Ritson Road to Highway 2, while at the other end the new route merged with it.

Thus, the first mileage demonstrated the principle of having each section connected to existing provincial highways, and to relieve the most critical areas of traffic congestion.

TABLE 1
ONTARIO MOTOR VEHICLE REGISTRATIONS

YEAR	PASSENGER	TRUCK, TRACTOR AND BUS	DUAL PURPOSE*	MOTOR CYCLE	TOTAL	DRIVERS' LICENCES
1903	178				178	
1904	535				535	
1905	553				553	
1910	4,230				4,230	
1915	42,346			4,174	46,520	5,322
1920	155,861	16,204		5,496	177,561	19,563
1925	303,736	34,690		3,748	342,174	33,740
1930	490,906	61,690	5,986	3,924	562,506	673,694
1935	489,610	67,590	2,370	4,506	564,076	707,457
1940	610,576	86,038	1,855	5,403	703,872	937,551
1945	555,461	100,234	1,279	5,745	662,719	971,852
1946	585,604	117,217	1,303	6,982	711,106	1,087,445
1947	645,252	140,930	1,294	9,471	796,947	1,144,291
1948	698,384	162,589	1,199	11,086	873,258	1,209,408
1949	771,709	183,598	1,035	13,027	969,369	1,278,584
1950	881,143	202,800	6,428	13,709	1,104,080	1,366,388
1951	958,082	225,271	8,275	13,470	1,205,098	1,401,538
1952	1,024,816	243,591	9,939	13,407	1,291,753	1,556,559
1953	1,117,175	261,923	13,707	13,314	1,406,119	1,656,259
1954	1,187,725	272,241	17,560	12,454	1,489,980	1,747,567
1955	1,292,133	287,942	25,457	12,321	1,617,853	1,856,845
1956	1,365,874	297,329	35,385	11,652	1,710,240	1,967,789
1957	1,431,438	304,568	45,971	11,522	1,793,499	2,088,551
1958	1,492,039	308,317	58,418	10,148	1,868,922	2,176,417
1959	1,573,365	316,272	74,014	10,086	1,973,737	2,270,246
1960	1,640,346	320,190	92,587	9,361	2,062,484	2,355,567
1961	1,686,149	322,882	108,295	8,944	2,126,270	2,414,615
1962	1,718,413	329,706	121,706	7,323	2,177,148	2,469,425
1963	1,790,788	333,701	136,090	7,741	2,268,320	2,555,015
1964	1,877,443	342,357	151,085	10,334	2,381,219	2,694,023
1965	1,976,625	352,914	163,071	24,070	2,516,680	2,739,138
1966	2,063,754	370,026	171,735	37,959	2,643,474	2,821,648
1967	2,134,287	381,081	178,057	42,941	2,736,366	3,004,654
1968	2,237,298	385,242	187,618	47,826	2,857,984	3,128,509

Source: Ontario Department of Transport
*Since 1950, station wagons have been classified as dual purpose

A term used by highway engineers to denote the amount of traffic a particular class of road can be expected to handle adequately is "practical working capacity". For two-lane highways, such as Highway 2, this rating had been set at 5,200 vehicles (composed of 90% passenger cars and 10% trucks), in a 24-hour period. But that was assuming the alignment, grades and sight distances were excellent— which they were not on Highway 2 between Toronto and Oshawa. And due to the many transports travelling regularly between Toronto and Montreal, the breakdown between passenger vehicles and trucks was nowhere near the 90:10 ratio.

The 8,000 vehicles a day using this section of Highway 2 in 1946 were far beyond what the road could comfortably accommodate. And the traffic was steadily increasing. By 1948, the year after the opening of the corresponding section of Highway 401, the daily volume over that new mileage averaged 8,500 vehicles. That same year the average daily traffic on the parallel section of Highway 2 dropped to 3,750. In 1949 the total for the two sections leaped to 15,500 vehicles a day. Thus, although the Highway 401 section in itself doubtless generated more travel, traffic would have hopelessly plugged this section of Highway 2 if Highway 401 had not been constructed.

The trans-provincial route in time has had a profound effect on both economic and social patterns of Ontario. The highway linked the major cities across the southern edge of the province, and encouraged location of new industry which could be served by a more flexible labour market able to accept jobs involving considerable daily travel. The following sections will consider some of the effects of the 401 and the complex procedures involved in its construction.

401, FORD AND ST. THOMAS

Locational factors

Quick, convenient access to major market areas has been a major factor in the location of plants near Highway 401. The freeway has attracted hundreds of industrial enterprises because it provides the most convenient link between large urban areas in Ontario, Quebec and the United States.

As a prime example, Ford Motor Company of Canada Limited chose a location a few miles south of Highway 401 for its $65 million assembly plant at Talbotville Royal, near St. Thomas.

The 1,331,318 square-foot plant went into production in 1967 and in 1969 employed over 2,500 people from more than 42 municipalities

in nine southwest Ontario counties. Most lived in London in Middlesex County, and St. Thomas in Elgin County, but others were drawn
from such communities as Woodstock and Tillsonburg in Oxford
County and Simcoe and Delhi in Norfolk County.

At the end of 1968, on the first anniversary of production, the year's
payroll totalled $13,068,500. Production materials purchased from
Canadian suppliers came to just under $27,000,000.

Ford's Canadian president, Karl E. Scott, in 1968 described the
region bisected by Highway 401 as being part of a "megalopolis
stretching from Quebec City to Chicago, with similarities to the eastern
seaboard of the continent.

"The Canadian sector of this corridor, which stretches from Quebec
to Windsor, is a 50 by 700-mile strip comprising less than 1% of
the area of Canada, yet containing 62% of the population, 84%
of the industrial activity and 41% of the commercial agriculture",
said Scott. "For a Canadian automobile company to locate an
assembly plant outside this corridor at the present time would not
make economic sense."

(Some auto assembly plants are located in Canada away from the
main market areas, but there are usually special considerations involved.
The most notable example is the Volvo plant in Nova Scotia. The
Swedish-based firm established its plant at Halifax because of the
availability of labour at acceptable wage rates, plus tax concessions
offered by the provincial government.)

What led up to Ford's decision to locate in the St. Thomas area?
The answer is simply economics. Scott emphasized this point:

"Nobody should be so misguided as to assume that industrial plants
are built in order to appease government economic planners or to
win special influence in any one locality. They are built in the
expectation of growth of markets in main streets all across Canada,
and in doing so, draw reciprocal benefits from the national growth
which reflect at the local plant community level."

"Investment decisions in the automotive industry are not the result
of any hasty action", said W. P. Mitchell, another Ford executive.
The time elapsing from conceptual planning to production might span
a decade or more, and it required at least four years from the time a
decision to locate was made before a plant could come close to full,
two-shift production. Preceding this, studies of market areas were
required to ensure that a plant was strategically located between the
sources of major components it assembled, and the geographic market
for which the most of the finished products were destined.

Mitchell said that the prospective general area must have a labour force adequate in terms of quantity and quality. This was a factor which eliminated many areas. The plant also required ample land with essential services, such as water, as a large, two-shift plant could use up to 1.5 million gallons a day. There must also be road and rail access, supplies of power and facilities for waste disposal. For its assembly plants, Ford required sites of at least 400 acres.

Mitchell noted that the expense of land represented a very small part of the total cost of a plant—traditionally only 1% to 2%— hence it was usually not a critical factor in the investment decision. He added:

> "Another aspect to be considered is the availability of a construction industry capable of erecting a plant to our requirements, and being available on a continuing basis for extensions, re-arrangements or major repairs. This factor has an important bearing on construction costs. The lack of these skills in a region would involve prohibitive costs if there were temporary importing of capable personnel."

In its search for a new location, Ford narrowed the choice to a southern Ontario centre which could provide an adequate labour force, a suitable site, and transportation systems with ready access to the United States market.

After many studies, St. Thomas was selected. It could provide the four essential services of water, natural gas, rail and highway transportation, as well as a large labour force trained in an industrial environment.

In addition to evaluating economic considerations, Ford also looked at the general social environment of the region. The company considered the attitudes of the public and local government officials towards business, the general character and civic reputation of the locale, the adequacy of hospitals and schools.

Scott pointed out that any plant built in a semi-rural area, but with the right transportation facilities, has a

> ". . . complex pattern of relationships, direct and indirect, with both neighbouring and distant communities. One example of this is that a person no longer needs to live, work or play within the boundaries of one municipality. Our age is one of rapid travel, super-highways and broadening horizons. Our homes may be in one municipality, we may work in another, shop in another and spend our leisure time in yet another. On the basis of our own experience in Oakville [Ford's head office in Canada] we know that it is nothing out of the ordinary for a person to travel 50 miles or more to his place of work in no more time than it takes within a big city to get from a suburb to the downtown area."

Ford of Canada, Ltd.

Clamshell carriers, rather than traditional floor conveyors, transport vehicles along the trim lines and part of the chassis line at the St. Thomas assembly plant of Ford of Canada.

Another factor is government programmes directed at regional planning. Tax concessions and grants will influence industry's decisions on the location of new facilities.

The Automotive Trade Agreement between Canada and the United States, which went into effect in 1965, was also important to Ford's planning.

The trade pact, allowing the free flow of automotive products across the border without tariffs or other impediments, broadened the potential for the St. Thomas facilities and gave the plant access to the entire North American market.

Although the car industry in Canada had already grown by 40% from 1962 to 1964, making it imperative that Ford construct a new plant, the introduction of the auto pact brought about a major change in the capacity planning studies being carried out.

Ford officials recognized that the new plant should be designed to recognize that much of its production would be shipped to the United States. Realignment of facilities, spurred by the auto pact, meant fewer Canadian models but larger production runs. The fall in the number of models permitted a significant decline in assembly costs, and made possible price levels equal to the best of the U.S. plants. The St. Thomas operation assembles only cars.

Company-community relationships

Professor E. G. Pleva, former head of the Department of Geography at the University of Western Ontario in London, has studied the Ford operation at St. Thomas and feels it is setting a pattern for future development throughout North America. In 1968, he described the industrial complex as creating a "ripple effect" by reaching out to adjoining communities for workers, and feels the establishment of large plants in generally rural areas may solve some of the big cities' problems involving urban sprawl and loss of identity on the part of the individual.

"Good roads and other facilities," he adds, "make it possible for workers to live in towns in which they are already established. This gives them the feeling of belonging to a community."

However, the construction of major plants in rural areas raises the question of the ability of small community governments to cope with this new industrial age.

Professor Pleva has suggested that the mobility of today's worker and the establishment of plants such as the Ford complex at St. Thomas have outgrown local government as it exists and the answer lies in larger, regional jurisdictions, a system now being introduced by the Ontario government. He added:

"When 84% of the population in the corridor are gainfully employed in manufacturing, when mobility has so changed that people can move tremendous distances to get to a plant, the result is confusion. Ten or twelve political jurisdictions may affect one modern family. Mobility has resulted in changing homes, changing jobs and changing cities. . . .

"According to the planners involved in the selection of this area as the site of the giant new plant, the established centres were to be the growth points—places like St. Thomas and London.. This was fine in theory, but when things were finally established, the township with the big plant got the taxes, and the established towns, nearly thirty of them, got the dormitory costs—education, fire protection, police protection, road maintenance, garbage disposal. Ford claimed that the assembly plant employed comparatively few people compared to the suppliers of cars and parts, etc.

which are scattered about the country, and that as far as they were concerned industrial location is a geography of costs and it is possible from one plant to supply a whole nation and a whole continent. Their choice was based on this consideration. Any problems with dormitory towns, misallocation of taxes, etc. were for the consideration of regional government."

Ford takes the overall view on corporate-community affairs, feeling it must look at the country's whole economy in fulfilling its obligations to both shareholders and the nation. Scott maintains that "if something other than economics is ever allowed to dictate the location of a plant site, then the economy as a whole will be endangered."

Looking at the general picture of corporate responsibility, he offered this description:

"Across Canada, thousands of small manufacturers bring the benefits of far-off big new plant developments to local communities. This is not difficult to understand if you think of an automotive assembly plant as being at the end of a series of trails that begin at an iron ore mine at Steep Rock, Ontario, or at the height of land on the Quebec-Labrador border. The ore mined at these locations is processed into iron and steel at mills in Ontario, Quebec and the United States and then the trails begin to resemble a spider's web as the metals find their way to the hundreds of stamping plants, tool and die shops, wire manufacturers and metal product plants across Canada which manufacture the thousands of components that go into a car or truck. Other trails lead from the textile mills of Quebec, and from the asbestos and aluminum industries of that province; from plastics and rubber manufacturing plants; from paint and chemical producers; from manufacturers of electrical components."

Scott said he fully realized that the precise location of a major plant was also of vital importance to host communities. "I also realize that if patterns change, if the markets catered to by an industry do not continue to grow, serious disruption in the economy and life of a host community can result. In the history of Canada there have been many examples of boom-or-bust towns, in particular those associated with mining, textile or lumbering enterprises", he stated.

Scott, who favours regional government assistance to small towns and villages facing industrial expansion, asked: "What can a community do to help prevent that industrial necklace from turning into a millstone around its neck?" His answer:

"It may seem that the influence of a local community on the economy as a whole can at best be only slight. However, by banding together in geographically economic blocs, rational economic planning on the scale that is required for an area such as this can be achieved and greater influence wielded.

"The narrow viewpoint that what is good for the town is good for the region, is good for the province, is good for the country, must be reversed. We must think first of what is good for Canada as a whole and then the local rewards will fall into place. In my opinion groups of counties should work together on regional problems to the mutual benefit."

Scott praised the leadership toward regional economic planning being given by the Regional Economic Development Council of Ontario, an organization divorced from politics. Ford of Canada located its St. Thomas plant in the Lake Erie Region of the Council, one of 10 into which Ontario is divided. The Lake Erie Region, composed of Elgin, Middlesex, Norfolk and Oxford.counties, contains 71 municipalities and has a population of about 500,000 within its 3,000 square miles.

Scott said he believed the duties of businessmen and government were intermingled, and offered two basic requirements of good corporate citizenship.

First, industry must avoid industrial paternalism and must not assume a dominating position in a community. It must also accept, commensurate with its size, responsibilities as a corporate citizen.

Second, industry must do what it can to ensure that communities share equitably in the benefits to be derived from its location in a certain area.

INDUSTRY AND KITCHENER

No longer is the large city so attractive to many industries, and the density of development within the large city is not so critical. Industries that perhaps at one time were attracted to a metropolitan area like Toronto are now freed to move to a less congested city, where a highly skilled labour force exists, where the cost of living is lower and where cheaper land is available for industrial sites.

Kitchener, in the 10 to 15 years prior to 1958, had experienced very little new industrial development. The oldest industrial establishments in the city had developed largely unplanned near what is now the centre of the city. Because of early industrial dependence on the railroad as a means of transportation, this old industrial district settled on both sides of the Canadian National Railways' main line. There were minimal zoning regulations as to site coverage or type of industry. As a result, many buildings covered almost 100% of the site

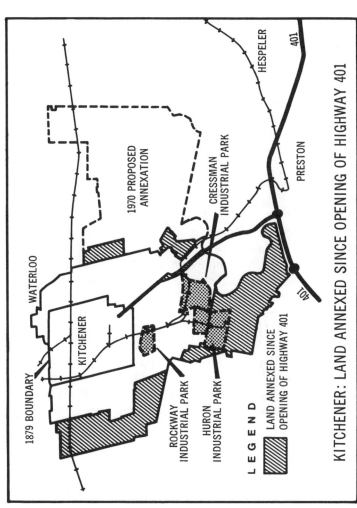

1879 BOUNDARY

WATERLOO

KITCHENER

1970 PROPOSED
ANNEXATION

ROCKWAY
INDUSTRIAL PARK

HURON
INDUSTRIAL PARK

CRESSMAN
INDUSTRIAL PARK

HESPELER

401

PRESTON

401

L E G E N D

LAND ANNEXED SINCE
OPENING OF HIGHWAY 401

KITCHENER: LAND ANNEXED SINCE OPENING OF HIGHWAY 401

Department of Planning, City of Kitchener

with very little room left for parking. Most of the plants were several storey, loft-type buildings, now obsolete for many modern industrial practices.

Industrialists were no longer attracted to a city which could offer only raw sites; they demanded serviced sites at competitive prices which could only be supplied by a public corporation. Kitchener, at this period in its history, was fast becoming a dormitory community with a large imbalance between its residential and industrial assessment. Land for industrial development in Kitchener was, prior to 1958, acquired by private enterprise and on a hit-and-miss basis. Since that time the city has purchased and developed land for industrial purposes: in effect, it was forced into the land development business. The city, for example, bought several farms to the south and west and over several years graded and serviced the land.

To avoid the pressures of congestion and decay associated with high central densities of population, many businesses located outside the central area of Kitchener. Several industrial parks were established to the south and west, nearest the best transportation access to Highway 401. The improved highway facilities encouraged the spread of many manufacturing functions beyond the old city and outside the once dominant influence of the railway lines. Nevertheless, all these industries were still within close contact with the city for the many benefits that still accrue to a downtown area. The new industrial parks were soon occupied by such companies as Westinghouse, Electrohome, Budd, Crown Zellerbach, Greb, and B.F. Goodrich.

The question is whether the industrial parks would have been sucessful without Highway 401. It would appear that few economic forces have been more influential in modifying the pattern of industrial location in recent years. In the period 1964-1969 the areas annexed by Kitchener have attracted 66 diverse companies, employing 3,800 people.

METRO TORONTO BYPASS

Before the last section of the 401 was opened, the increasing popularity of the Toronto bypass section made reconstruction and expansion essential.

The original four lanes across the top of Toronto had been designed to handle easily 35,000 vehicles a day, and were mainly intended for drivers wanting to bypass the city. But things did not work out that way, for residential sub-divisions and industrial developments were soon attracted to the new route. Commuters began to use the 401 as a

kwood Survey Corp., Ltd.

Lockwood Survey Corp., Ltd.

Before and after 401. Only 7% of the gainfully occupied population of Ontario works in agriculture, and thousands of acres of agricultural land in urban areas are covered by new buildings every year. These photographs were taken in 1947 and 1969 near the intersection of the 401 and Highway 400, Toronto.

regular route to work. By early 1961, sections of the bypass were carrying up to 70,000 vehicles daily. Two years later the traffic count had climbed to 85,000 a day, higher during peak periods. Checks showed as many as 4,000 cars and trucks travelled in one direction within one hour—more vehicles than some sections of provincial highways carried during a month or more. (Practical capacity for a four-lane highway of the 401 type at that time was generally considered to be 48,000 vehicles a day.) As the volume mounted, travel speeds were reduced but there were more accidents, particularly rear-end collisions. In 1961, property damage alone for accidents on the 401 amounted to $400,000.

The DHO recommended replacing the most heavily-travelled portions of the bypass with a 12-lane design. The plan called for six central lanes, three in each direction, to handle through traffic, flanked by six outer lanes (again three in each direction) for local traffic. Transfer lanes would allow high-speed interchange of vehicles from the two groupings.

As soon as the plan was announced in 1962 there were outcries from various parts of the province that too much money was being spent for roads in the Toronto area, that these same funds could be more judiciously applied to other road problems.

Toronto area residents were well aware of the benefits they had derived from the 401's existence. Metro Chairman Fred Gardiner declared in 1961 that the freeway had been "worth $200 million" in providing fast, economical travel and in moving metropolitan development northward.

At the same time, the area in a 30-mile radius of Toronto's City Hall, holding approximately 20% of the country's population and almost one-third of the province's people, was making a sizeable contribution to Canada's corporate wealth. In Metro Toronto alone, population was increasing by 1,000 residents a week. An adequate transportation network was essential—and an expanded 401 was a vital link.

It was also evident that the area municipalities representing Metro Toronto, hard-pressed to find funds for local road improvements, could not finance a route parallel to the 401. Even had Metro decided to build a major east-west route to relieve pressure on the freeway, the DHO would have had to pay 50% of the cost (the rate of subsidy provided for such projects).

Because of such considerations, the Ontario government approved the large reconstruction project, which was designed to handle 164,000 vehicles daily. The limits of the 17-mile-long job were Highway 48

Lockwood Survey Corp., Ltd.

Department of Highways, Ontario

This is the Avenue Road interchange in the Toronto section of 401. The upper photograph, taken in the early 1960's, shows congestion which had developed along the four-lane route. In the lower photograph (1964), major changes had produced a 12-lane highway with longer bridges.

(Markham Road) to the east and Islington Avenue, later extended to Highway 27, to the west.

At that time, the basic 12-lane design was believed to be the largest highway facility of its type ever built as a complete unit anywhere in the world. The size of the undertaking was a tremendous challenge to engineering staff both inside and outside the DHO, and spurred designers to conceive and work out applications for bold approaches to brand-new problems.

For bridge designers it was the start of a succession of major advances —some already thought out but, up to that point, lacking on-site application. An early accomplishment was the introduction of a three-span structure stretching more than 340 feet across the freeway. This design eliminated shoulder piers and the driving hazard they constituted. In its first use, the Avenue Road interchange, and in other spans that followed, the only piers within the travelled portion of the right-of-way were placed within medians and sealed off from drivers by double steel beam guardrails. The overall effect has been a much greater "see-through" safety factor for motorists.

One of the largest and most complicated crossings on the freeway is the three-level Spadina Interchange. This maze of 24 bridges provides links with Dufferin Street and the vast Yorkdale Shopping Centre.

To produce the Spadina complex, which in itself contains 25 miles of pavement, there had to be a new approach to structural design. There were three main problems. First, two dozen structures had to fit into a restricted area. Second, because of road alignment and other factors, pier supports had to be squeezed into a tight space. Third, the bridges would require pronounced horizontal and vertical curvatures.

To find the answers, the DHO looked for precedents on this type of design. This meant sifting through mounds of engineering data from all around the globe, much printed in a variety of languages. Fortunately, the department had recruited from many countries following the Second World War and technically-trained people in the bridge design group alone represented over 20 European nations.

From this mass of data and other research efforts, engineers decided on a "continuous, post-tensioned concrete structure". In effect, this meant building the bridge sections in place, rather than transporting beams from distant pre-casting plants and then linking them together.

The design offered a light, clean-lined look, suitably strong and aesthetically pleasing, which relieved the congested appearance of the intricate interchange. At the same time the construction choice was well-suited to the sharp horizontal and vertical curves required in the

Department of Highways, Ontario

Concrete's magnetic attraction for people is well illustrated by this aerial view of the Spadina interchange, Toronto. Yorkdale Shopping Centre is in the foreground.

engineering. The dramatic simplicity of the interchange—which was built rapidly and was economical despite its complexity—helped to earn it a National Design Council award in 1967.

The directional signs and the lighting used on the reconstruction offer other examples of new approaches to highway construction. Easy recognition of route changes is an important safety factor on high-speed freeways, and to ensure maximum identification, the DHO selected two colours to distinguish between the two types of driving lanes. Directional signs on the outside or "collector-distributor" lanes are blue, while signs marking the set of inner or "core" lanes are green. In addition, the overhead signs have permanent catwalks so that they can be maintained without pedestrians being on any of the roadways.

The DHO worked with a consulting firm to establish a new concept for freeway illumination, resulting in "daylight" driving conditions at night.

Metro Toronto's population keeps growing to meet the bypass expansion. By 1969, it was well past the 2,000,000 mark, an increase of nearly a million since the first freeway section opened in 1947. By the year 2001, it is estimated that 4,500,000 people will be living and working in the area.

Reflecting this population boom, 120,000 vehicles a day swarmed over the first 13 miles of the reconstructed bypass when it opened in 1968, making it one of the most heavily-travelled highways in the world.

It is doubtful that the word "bypass" will ever again be applied to a new route built to skirt the Metro Toronto area. As soon as it became evident that the 401 had become a main route for Toronto region residents, planners looked to the possibility of utilizing Highway 7, a few miles to the north, as the new east-west bypass route. But already residential and industrial development has spanned much of the distance between the two highways and it is almost certain that Highway 7 will also be enveloped in time.

EVOLVING PATTERNS OF TRAVEL

The service centres

Service centres, those often welcome rest and food stops and refuelling centres found about every 40 miles along the route, were not provided along any section of the freeway until 1961. This was government policy based on the feeling that the freeway should generate as much business as possible for local communities, thus overcoming some of the criticism of freeway detractors who feared the loss of customers. Therefore, the Conservative government of Leslie Frost throughout the 1950's ruled that drivers should be directed off the freeway to adjoining communities for fuel, food and rest.

In addition, by barring commercial establishments along 401, the aesthetic qualities of the route would be maintained. (Ontario was a North American pioneer in eliminating advertising billboards from the right-of-way of King's Highways.)

However, public opinion brought about a change in the service policy. As soon as it became possible to drive continuously on the freeway for upwards of 100 miles, motorists began to demand gasoline outlets and other service facilities. In 1960, there was a 160-mile uninterrupted stretch of driving between Kitchener and Belleville. Not one rest area marked the route. The public clamour became an uproar. Drivers looking for more convenience and wanting to maintain a good travelling time did not want to leave the freeway to become involved in local traffic conditions.

The upshot was a change in the government's position. Service centres began to appear at 40- to 50-mile intervals, built as matching pairs, one on either side of the route. (Because there were sometimes site problems—unsuitable land or drainage difficulties—the "pairs" were not always exactly opposite.) Also, with the substantial investment in each centre, an agreement exists between Ontario and Quebec not to allow a centre within 25 miles of each side of the provincial border.

Department of Highways, Ontario

Service centres (this one is near Kingston) provide 24-hour restaurant and vehicle service. They are located as matched pairs on each side of the highway at 40- to 50-mile intervals.

The government decided that oil companies operating the service stations should be responsible for both the stations and restaurant facilities, a formula that has been followed ever since.

Basis of the bidding for the sites was the percentage of gross revenue each company was prepared to pay the government, the highest bid being the winning one. In addition, a fixed yearly ground rental was established for each location. There were 25-year leases to ensure that the department would recover its outlay in providing the land, grading the site, constructing entrance and exit lanes or ramps, supplying water and other such services.

One of the key stipulations was that each centre must remain open on a 24-hour basis every day of the year, with sufficient staff to provide a towing service, or at least to move vehicles to a safe area away from the traffic flow. This round-the-clock service was better than any service in existence for locations off the freeway.

Each oil company was free to come up with its own design, subject to final DHO sanction. As a result, each centre is different. Another provision was that each centre must set aside 150 square feet for an information booth, if desired. This would give local interests an opportunity to extol the virtues of their areas.

By the summer of 1962 the successful bidders had opened temporary facilities at the first eight sites, and by 1969 all 19 centres in the programme were operating.

Once the centres had opened, DHO inspectors visited each site at least twice a month to ensure that the operator was living up to the terms of the agreement. The visit included checking the restaurant menus, which had to offer an adequate variety of meals at reasonable prices.

It soon became apparent that the freeway service facilities were a major source of employment, in certain regions. During the peak summer months each centre employs up to 125 people during the three eight-hour shifts. Even in the middle of winter most centres have a staff of 50 to 75 employees.

So that the centres will contribute still more to the convenience of the motoring public, the DHO maintains large picnic areas adjoining each site and utilizing the same entrance and exit lanes.

Criticism and praise

One particularly vocal critic of the centres, a motel operator, proclaimed again and again before the first centre opened in his region that it would probably drive him out of business. However, a few years after the first centres opened, this same motel proprietor added many more units.

Sometimes other commercial enterprises were concerned about the effects of the freeway. Nor was this unease always limited to business activity. There were misgivings about the permanent disruption of road usage patterns. The severance of farms and other property, the disturbance during construction, and the inevitable dust were also subjects of criticism. Militant opposition delayed the start on some sections, often for a considerable time.

If the starting day of construction had been opposed, the official opening day for a section was usually just as eagerly awaited. It generally called for a round of regional rejoicing, and often a group of communities would band together for a collective celebration.

The passing of time showed how wrong, for the most part, the earlier fears had been. Area newspapers began publishing stories of various freeway benefits—the one cited most frequently was the new ease with which people could drive into the shopping and business sections of the bypassed communities and then find parking spaces. As a result business in particular, and communities in general, began to benefit from the 401. This pattern was repeated all along the route.

The freeway also brought benefits to the traveller. As one DHO official pointed out, 401 provided a saving in fuel consumption and vehicular wear and tear, as well as a reduction in personal injury and property damage accidents—the accident rate for the 401 is only about half what it is for the rest of Ontario's highway system, taken as a whole.

The safety features of the freeway also had the effect of enticing more women drivers to use the route. This new sense of confidence encouraged women to make longer and more frequent trips. As visitors from the country assaulted the department stores of the major cities, so in turn did the urban shoppers seek the benefits and bargains of the countryside. There were day-long forays into the farm belts; antique collectors searched out new finds, while others scouted for country-fresh produce.

Cultural activities were similarly vitalized. There is no doubt that the 401 was a boon to the Stratford Festival. This theatrical venture was launched in 1953 in a comparatively small community of 20,000 on the Avon River, 98 miles west of Toronto. Shakespearean productions, ballet, classical and jazz concerts and a variety of other attractions, drew critical acclaim and the Festival's popularity as summertime entertainment grew rapidly. Because of Stratford's accessibility via the freeway, theatre-goers from metropolitan areas as far away as Buffalo and Detroit became regular patrons. Similarly, Toronto theatres have attracted subscribers from as far as Kingston, 198 miles to the east.

The freeway also became a high road to advanced learning. Students from rural areas could continue to live at home and use the 401 to attend schools or universities. Parents thus benefited by saving the living-away expenses for sons and daughters, while retaining extra hands for farm chores.

Recreation

Holiday habits quickly changed. In the Metro Toronto region, for example, the freeway's ability to cut travel time revived interest in the Bay of Quinte and The Thousand Islands as vacation areas and sites for summer homes. The northeast Ontario resort regions also boomed. On Friday afternoons and evenings throughout the year the 401 becomes an avenue of escape for thousands of city and suburban dwellers to cottages, lakes and beaches in the summer; the ski slopes and snowmobile runs in the winter. In addition to the regular cottage and camping crowd, the freeway also attracts many United States tourists, to the benefit of resort operators in many areas of the province.

Ontario and Quebec travel ties have been strengthened by the 401. In line with this development, the freeway has become a main point in the promotion of the Ontario portion of the "Heritage Highways" route linking the two provinces. Beginning at Niagara Falls, the journey touches Hamilton, Toronto, Ottawa, Montreal and Quebec City, ending at Percé on the tip of the Gaspé Peninsula.

Characteristically, the freeway makes it easy for people to join this pioneer travel trail from almost any direction. There are also many side trips to choose from, including Sainte-Marie-among-the-Hurons and the Martyrs' Shrine at Midland (site of the martyrdom of Jesuit priests in 1649), or Fort Henry at Kingston, built in 1836.

One of the biggest attractions is Upper Canada Village, on Highway 2 east of Morrisburg. Reconstruction of a pioneer community, using in part historic buildings from land to be flooded by the St. Lawrence seaway project, commenced before the freeway had been completed in that part of eastern Ontario. Without the 401, or its equivalent, investment on the scale required to operate this major tourist attraction would not have made economic sense. Officials say that at least 75% of the visitors reach the village via the freeway.

Toronto was often the focal point for measuring the time-saving factors of each new section of the freeway put into service.

When an 18-mile section was opened in the western part of the province in October, 1963, the *Chatham Daily News* headlined the fact with: "Kent County 1¼ Hours Closer to Toronto". The *Windsor Star* used this banner: "Makes Chatham 1½ hours Closer to Toronto".

Absence of Tolls

In fact, as far back as the late 1950's enough mileage was in use to establish beyond all doubt the unique economic and social value of the freeway, so much so that some people became impatient with the construction schedules. They wanted them speeded up by charging tolls.

The result was the Ontario government's appointment of a Select Committee on Toll Roads and Highway Financing. This group was to study the possible application of tolls on both the 401 and future multi-lane, controlled-access routes. After a detailed investigation the committee came out strongly against the construction or operation of toll roads. They felt that the 401, and all heavily-travelled major highways, generate enough gains in a relatively short period to more than offset the initial construction costs.

An editorial writer for the *Peterborough Examiner* summed up the toll question this way:

". . . Yet in Ontario, despite the astronomical cost, there are some of the finest modern roads available to the motorist. Moreover, the building of them has been assumed as a public liability and, unlike the autobahns and the turnpikes, they are not subject to tolls. In this, Ontario has shown much good sense: it would have been easy to retire the cost of these highways by charging tolls but every resident of Ontario derives benefit from such a means of communication. Goods can be transported at equitable cost when the capital expense involved is assumed by the community. . . ."

And another newspaper, *The Stratford Beacon-Herald*, had this to say about the freeway:

". . . Highway 401 has speeded up travel throughout the province and has provided the more than 3,000,000 people who live close to it with an efficient means of cutting down distances to manageable proportions. No longer is it a formidable chore to chop one's way through traffic to Toronto . . . and a trip to Ottawa now takes at least two hours less than it used to. . . ."

In the final paragraph, the editorial commented:

"With the high density of population along the entire length of 401, the highway has brought prosperity to this area far in excess of the cost of building it. . . . The absence of toll charges . . . make Ontario residents the most fortunate of people in their travels."

PLANNING AND DESIGNING THE 401

Priorities: 401 versus 400

Why, with the first section of the 401 open in 1947, did it take nearly 20 years to close the final gap in the 510-mile route?

Part of the answer lies in the post-war shortage of materials and the serious backlog of provincial highway projects which was created by the war. And, for good measure, there was a post-war boom of automobiles and trucks seeking new and improved roads.

The way in which the people of Ontario began to use cars after the war went far beyond any reasonable forecast. With gas rationing a thing of the past, and automobile assembly lines rolling at top speed to fulfil peacetime needs, it seemed that everyone was out on the roads. There were waiting lists at auto dealers and you did not quibble about colour combinations or fancy extras.

Between 1946 and 1952, motor vehicle registration shot up an amazing 75%. During those same years, the trucking industry in

REGULATIONS

The popularity of the freeway has worried some officials. They fear Highway 401 might be reduced to a corridor of manufacturing plants, warehouses and shopping centres. This may be a danger for some roads without controlled access, but it is not a threat to the 401 as long as existing legislation remains in force governing construction along the strand of the freeway.

Under Ontario's Highway Improvement Act, the province through the Department of Highways maintains control on what may be built near a controlled access highway, such as the Macdonald-Cartier Freeway. On a controlled access route, the department decides the interchange points, and no other crossings are allowed.

A permit is required from the provincial Minister of Highways before construction can be carried out within defined distances of controlled access highways. For instance, approval must be received before anyone shall "place, erect or alter any building, fence, gasoline pump or any other structure on any road within 150 feet of any limit of a controlled-access highway or within 1,300 feet of the centre point of an intersection."

Again, permission must be received for construction within half a mile of controlled route interchanges for such projects as shopping centres, stadiums, fair grounds, race tracks, drive-in theatres, "or for any purpose that causes persons to congregate in large numbers". Regulations also cover the use of any private road, entrance-way, gate or other structure or facility as means of access to a controlled-access highway.

Regulations are also in force controlling the size of advertising signs within a one-quarter-mile of any limit of these routes. Where approval is granted, the sign must not exceed two feet by one foot.

Ontario became a much more significant factor in the transportation field. By 1950 there were 200,000 commercial vehicles registered in the province, more than double the 1945 figure.

With both the driving boom and the backlog of highway work across the province, it would have been illogical and, indeed, political suicide for any government to have concentrated all its major road-building efforts on the 401 alone, no matter how important it was.

Instead, the Ontario government, through the DHO, allocated yearly expenditures on the provincial highway system on the basis of priorities, regardless of the class of highway concerned. With new highways the policy then, as now, was to do as much work as possible with what funds remained in the budget.

In the immediate post-war period, the 401 took a back seat to another controlled-access route, a new north-south freeway between Toronto and Barrie. It was designed to relieve the intense congestion on Highway 11, the primary road link to northern Ontario and all the cottage and resort country en route.

Department of Highways, Ontario

Increased sizes of equipment made it possible to carry out the major clearing and grading jobs required in providing the wide right-of-way for 401. This is a section in the Toronto bypass, facing west from Keele Street, under construction in June, 1951.

Work on this new two-lane highway, the 400, started in 1946. By the summer of 1952 the first 50 miles had been completed as far as Barrie.

The heavy annual expenditures for the 400 during those six years did not leave much for the Macdonald-Cartier. But it was making progress. In 1951 and 1952 work was underway on: a 25-mile section east of Windsor; a 32-mile stretch between London and Woodstock; a section east from Oshawa; a short section north of Kingston, and parts of the Toronto bypass.

At that time, department reports still referred to the bypass as an "interceptor" route to draw off through traffic that would otherwise have had to fight its way through heavy traffic in and around Toronto, further to the south. Those were the days before the bypass route had attracted Metro Toronto northward.

Before the end of 1952, the DHO opened the bypass west of Yonge Street (Highway 11) for seven miles. The route now reached an intersection with the north-south Weston Road and, together with Highway 27, a north-south route to the west, provided a choice of links to the QEW.

Construction might have gone faster, but once again war placed a strain on manpower and resources. This time it was Canada's commitment to Korea, which, until the armistice of July, 1953, hampered construction at home.

The route

Before discussing construction progress on the 401, let us go back to the late 1930's and take a look at the planning of the route.

The planners knew that the 20-mile corridor through which they plotted the proposed route at that time contained more than 3 million people—over 65% of Ontario's population. And the percentage of vehicle ownership was even higher.

Statistics of that kind, and other related information, were featured in an address to the Royal Canadian Institute at the University of Toronto in February, 1954. The speaker was William J. Fulton, the DHO's chief of surveys, who later became head of the planning and design branch and then deputy minister.

He explained how origin-destination surveys, followed by aerial photography, were used to fix the route. These surveys were conducted at various check points between Windsor and the Quebec border. Drivers were stopped and asked where they had begun their trip, their destination and the route they planned to follow. Answers were obtained

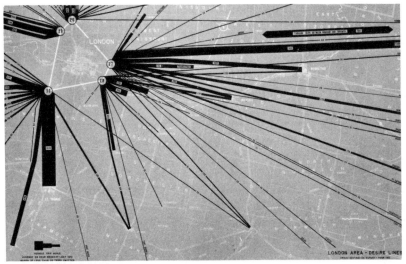

Department of Highways, Ontario

Highway travel "desire" lines were established by the Ontario Department of Highways from origin-destination surveys. This one was in the London area in 1951. Note how, except for the short inter-city movement between London and St. Thomas, the trips between London and Toronto overshadow all others.

from over 375,000 drivers, an undertaking which kept department crews busy for three summers.

Once this mass of data had been analysed, the next step was to plot "major desire lines of travel" uncovered by the surveys; that is, the route most people would take between any two given points if such a route existed.

A proposed route location for a section of the 401 between two such terminal points was then projected onto aerial photographs. With a line determined in this way, engineering survey crews ran or established this same line on the ground.

Overlapping vertical aerial photographs were matched stereoscopically so a photogrammetrist could observe in three dimensions and thus determine the topography. The photogrammetrists, by studying these photographs (taken more than a mile above ground), could then trace drainage patterns, distinguish cultivated land from pasture and, by identifying tree types, tell both the class and condition of the soil. The mass of detail they extracted was such that drawings could be prepared to show differences of as little as half a foot in land elevation.

In their photographic studies, these experts always looked for a height of land or ridge upon which to locate their proposed line, thereby greatly simplifying drainage problems and bringing other advantages. For example, it was easier and cheaper to fit structures for intersecting roads into the total design.

Watchful eyes looked for eskers—ridges or mounds of sand and gravel. These glacial formations are as ideal for road foundations as they are for drainage.

(Professor Pleva credited this feature of the design with saving millions of dollars in construction costs. As a bonus, the high ground provides drivers with a scenic view.)

In addition to geographic planning, social and economic objectives had to be considered. A route should cause the minimum of disturbance to local inhabitants, and at the same time offer the greatest service in moving traffic safely and efficiently.

With these and other factors in mind, the photogrammetrists drew a line on the photographs that would best fit the topography while staying as close as posible to the major desire line.

Wherever practical, in rural areas the 401 was located between two concession roads, and property requirements were obtained by taking land from the back acreage of adjoining farms. Again, where possible, equal amounts of land were taken from each farm for each 300-foot right-of-way. Then, as now, the DHO's prime concern was to keep to a minimum the severing of property, a policy that saved money and improved public relations.

After proposed lines for the route had been established, they were marked on maps drawn to a scale of one-inch-to-400 feet. These maps, which covered an area of about one mile each side of the centre line, showed such features as buildings, fences, power lines, railways, roads, wooded areas and streams.

Next, location survey parties were sent out to fix the centre line on the ground. Where a final choice had still to be made, this sometimes involved setting two lines.

At this point land titles were searched in local registry offices, municipal assessment records, and other sources. This information on property ownership and boundary lines was indicated on the one-inch-to-400-feet maps. However, at that stage there was no attempt to acquire any property.

Based on the knowledge obtained by the ground survey party, a new plan was drawn on a scale of one-inch-to-100 feet. This covered an area of about 300 feet on each side of the centre line. A profile and cross sections were taken along this same line, and additional surveys were made at river crossings, intersecting roads and any other location where greater detail was needed for the design of bridges, interchanges and other structures.

Detailed design

Only after completion of these vital intermediate steps did actual design work on the highway begin and property requirements become precisely determined. Estimates were made of the amount of earth and rock to be moved, as well as the construction materials and supplies that would be required.

Work also proceeded on several kinds of drawings that had to be made by various department groups. Each mile of the route took thousands of man-hours of art work.

The functional planning group, in producing their drawings, considered the effect on property of the road pattern they were recommending. They also had the task of detailing to others just how the road would function and the level of service it should provide.

Planning spokesmen spent considerable time explaining to the appropriate municipal officials and others the effects these functional plans might have. As a result of these discussions, the plans were frequently modified.

This approach has been departmental policy for many years for all proposed construction. In the case of Highway 401 there were many instances in which the planning of certain sections was held up because the DHO was unable to readily determine just what the municipalities concerned wanted. One of the main stumbling blocks was that communities were often unable to decide on the freeway connections that would best serve them. Sometimes they were uncertain about their own future road plans. But since both were matters of crucial importance in the planning stage, the department had to wait until agreement was reached.

With the location and design concept for a particular section of the 401 settled, the road design group began work on a second category of drawings, taking over where their functional planning associates had left off. Using the latter's plans, the engineers and technicians went into much greater detail to cover every aspect of design.

Because of the time required for functional planning (up to two years in some cases) much more knowledge was available for the next stage of design by the time they had finished. This meant changes. The discovery of poor soil conditions in a certain area might require alteration of the route. Perhaps certain property could not be purchased at a justifiable price, or perhaps it was simply not obtainable.

From the schematic drawings for the 401 interchanges and other components prepared by the functional planning group, road design personnel developed an exact geometrical design.

The design work was done in two distinct stages. The first, covering everything up to the completion of grading and drainage work, was usually in units of four to eight miles. For the second—the laying of granular base and paving—the sections ran to about eight miles, sometimes slightly longer. These plans included every detail of construction.

It is from these drawings that contractors bid the work when it was called for tender. And for each section of road called, prospective bidders obtained a folder containing up to 75 pages of such drawings.

In addition to the time spent on road design, the department's bridge design branch required not less than six months' intensive work for the structural detailing of even a relatively simple span.

In obtaining the necessary land for highway construction, the DHO's property section carried out surveys on both sides of the line of route to determine the exact limits of the property required. This produced the third and final major set of drawings. On the average, plans drawn for one mile of Highway 401 right-of-way and the adjoining land covered a sheet of paper 13 feet long and 14 inches wide.

Computers

There are no "instant" highways, as can be determined in this basic description of the major steps taken in producing the Macdonald-Cartier Freeway. It takes a full four years—give or take a few months—to plan and construct just one section of controlled-access route like the 401. This time element, one restriction on mileage that could be put in place each year, would have been even longer without the extensive use of computers. The DHO was a pioneer among North American highway jurisdictions in harnessing the computer for road design. Thousands of split-second calculations eliminated many of the repetitive and tedious mathematics common in engineering.

Computers were also applied to photogrammetry, soil investigation and analysis, land surveys, bridge and road design and, in particular, to analysis of the massive data gathered in origin-destination surveys.

One of the earliest land survey applications was computer calculation of the earth and rock quantities to be moved to obtain the desired road grade. This was an important factor in comparing the merits of one route over another.

ASSEMBLING THE SECTIONS

The detailed planning which preceded construction began to have tangible results. By August, 1956, you could drive on the 401 for the full 24-mile distance between the eastern and western limits of Metro Toronto. This section, combined with the previously placed eastern section, made it possible to continue through to Oshawa and beyond, to the junction with Highways 35 and 115 west of Newcastle, a distance of over 55 miles.

During that period there was a certain amount of scepticism among the Toronto newspapers about the route of the bypass. They asked how the DHO could possibly justify acquiring 300 feet of right-of-way across the breadth of Metro Toronto for the 23-foot pavements—a width one foot greater than the intitial pavement between Toronto and Oshawa.

One featured article on the highway emphasized, by means of large pictures, the verdant qualities of the route. But the magnetic qualities of the 401 soon become evident, surprising much of the public and even the professional planners. Almost overnight the Toronto bypass became a commuter's bonanza. The new ease of travel advanced the shift of population to the suburbs; housing came first, followed by industry. Almost overnight, it seemed, subdivisions of hundreds of homes sprang up on what had been farmland or largely vacant land adjacent to the bypass.

The next year, 1957, the freeway grew impressively, as three new four-lane sections totalling 70 miles were opened. And over most of that mileage there were 24-foot pavements.

One of these sections, in eastern Ontario, was the first phase of the 13-mile Kingston bypass—from a newly-located section of Highway 38, east to Highway 15. Less than a year later, the second phase of the bypass went into operation.

In the west, two sections went into service—a 26-mile portion between Windsor and Tilbury and a 38-mile stretch between Highway 4, southwest of London, and Highway 2 at Eastwood. Generally called the London bypass, this was the largest section of the 401 ever opened as one unit.

Several aspects of this particular section illustrate key principles followed by the DHO in planning, scheduling and constructing the freeway.

As well as relieving the strain on Highway 2, the new section was a boon for traffic originating both north and west of London, extending as far as Sarnia and the United States border crossing point of Point Huron, Michigan. The department also spent $200,000 to improve four miles of county road, providing a direct route between the 401 interchange at Wellington Road—one of London's main arteries—and Highway 2 near Lambeth. This improved road became King's Highway 135.

Aesthetics as well as cost savings were intermingled in the London section. The right-of-way was grass-seeded or sodded throughout and some 26,000 trees and shrubs were planted to heal construction scars.

In 1967, the Department of Highways, Ontario, issued a Highway Planning Study for Southwestern Ontario, one of 19 regional studies conducted by the department to establish a highway plan for the next 20 years. The study, carried out over two years by the Traffic and Planning Studies Division of the Planning Branch, included the Counties of Essex, Kent and Lambton, an area of 2,750 square miles.

Traffic information was obtained from counts taken in 1964 at key locations along the highway network and from two types of origin-destination surveys, one being a roadside interview and the other a telephone interview. The roadside survey recorded about 55,000 driver interviews. The telephone survey covered homes in rural areas and provided data on more than 20,000 vehicle trips.

Land use data for the area, in terms of the numerical distribution of population and employment by traffic zones, was supplied by the Community Planning Branch of the Ontario Department of Municipal Affairs. Predictions of 1985 traffic were obtained through electronic computing procedures that employed the land use data in conjunction with the 1964 traffic figures.

Analyses were made of the 1964 and estimated 1985 traffic patterns, to find the highway sections that would or might become insufficient within the 20-year planning period. Other comparisons were made to validate the accuracy of the traffic volume data and the origin-destination surveys.

Computer programmes were prepared to select routes either by the least time or the shortest distance between centres on various roads or theoretical networks. The trips from origin-destination tables were accumulated on the various links which form these minimum paths, to get the volumes on the roadway section.

Other conditions, including speeds of travel, location of roads, and quality of surface were varied to represent either actual or desirable conditions. Different lengths of trips were assigned separately to distinguish between roads with provincial significance and those considered to be local in function. Completely new links were inserted to test their effects on the rest of the network.

Motor vehicle travel, in terms of trip volumes from traffic zone to traffic zone, was plotted graphically on minimum time and minimum distance paths, using existing travel routes only. Then, assuming that such roads and highways were non-existent, a theoretical link network (the composite network) was developed to simulate the area's vehicle-trip

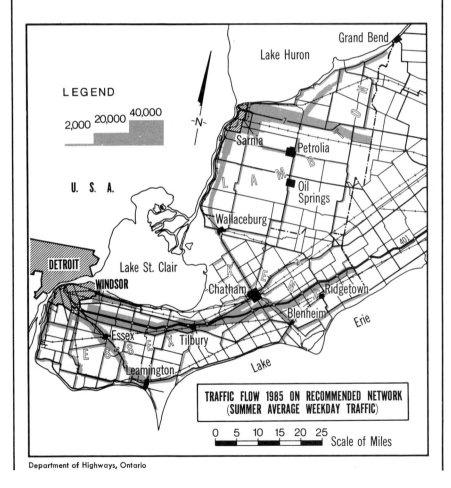

LEGEND

2,000 20,000 40,000

-N-

U. S. A.

DETROIT
WINDSOR
Lake St. Clair

Lake Huron
Grand Bend

Sarnia
Petrolia
Oil Springs
Wallaceburg

Chatham
Blenheim
Ridgetown
Erie

Essex
Tilbury
Leamington
Lake

TRAFFIC FLOW 1985 ON RECOMMENDED NETWORK (SUMMER AVERAGE WEEKDAY TRAFFIC)

0 5 10 15 20 25
Scale of Miles

Department of Highways, Ontario

distribution over the most direct paths. The latter technique revealed the true directional travel desires, by actual volume, in and through the study area.

The accompanying map showing projected 1985 traffic volumes by trip length and direction was among those prepared to assist in determining the level of service that each highway route should provide.

An examination of the travel desires indicated two main corridors which merited freeway service: the east-west route from the eastern boundary of the study area to Windsor; and the east-west corridor following the route of Highways 402, 7 and 22 from Sarnia to London. The former, the 401, in 1964 carried a daily average of 6,500 to 7,500 vehicles and the latter between 5,500 and 8,000 per day.

Most journeys in these two corridors are over 50 miles in length. The 401 will be adequate for the 1985 estimated traffic volume of 12,000 to 27,000 vehicles per day. (Average 1969 volume was about 8,200 vehicles per day).

The Sarnia to London corridor capacity analysis indicates that Highways 7 and 22 will become deficient during the 20-year planning period. The study, therefore, recommended that a freeway be constructed on a new alignment.

The traffic could be served equally well by locating the freeway either north or south of Highway 7, but if residential development east of Sarnia were considered then a route north of Highway 7 would be the more desirable. Such a route would also assist the flow of recreational traffic from the United States and Sarnia to the south shore of Lake Huron near Grand Bend. Four lanes would be required to meet daily 1985 traffic volumes estimated to be 11,000 to 28,000 vehicles.

In another part of this section, planners deliberately routed the 401 through more than a mile of swampland. The main purpose was to keep land severance to a minimum and to save money, since the area is prime farmland country. Although the Dorchester Swamp east of London presented major construction challenges, the problems were overcome. The maximum depth of the swamp was 30 feet and, on the average, the top 15 feet of organic matter was skimmed off and replaced with suitable fill. An extra six feet of material was placed above the road grade desired and this weight mass forced down the fill, an engineering technique known as "surcharging".

When each of the three 1957 sections was opened only the essential structures required to carry intersecting roads were in place. With the 38-mile London link, 15 structures had been completed, four more were under construction and nine were scheduled for the future. By doing this the DHO could stretch each year's budget and open the greatest possible mileage.

These three sections meant that 167 miles of the freeway were now in use. This was the equivalent of over 400 miles of standard King's Highway: the money spent on the London section alone would have built 150 miles of the other class.

In 1958, the opening of two adjoining sections in eastern Ontario meant a continuous 33-mile drive east from Brighton almost to Marysville, east of Belleville, where a connection was made with Highway 2.

At the Windsor end, the freeway was extended in the shape of a "Y" to provide a high-speed interchange for routes leading to either the Ambassador Bridge or the Windsor-Detroit Tunnel, the busiest border crossing points in Canada. The Toronto bypass also grew westward from Highway 27 to Highway 10. All told, the sections added that year came to just under 45 miles.

DHO policy was to build only two lanes at first instead of four, if this would adequately meet traffic needs for a reasonable length of time. A dozen or more sections were opened under this policy. One exception was the 18-mile section between Newcastle and Port Hope, opened in 1960. This is rolling countryside, for at this point the freeway skirts the southern fringe of the Northumberland Hills. Because of the heavy cuts and fills necessary to get the desired grade, it was cheaper to complete the grading for all four lanes in one operation. This also applied in areas requiring large quantities of rock removal.

Similarly, where practical, all four lanes were paved at the same time, even if only two lanes were actually required to carry the current traffic.

(A driver on the 401 might wonder why concrete predominates as the primary pavement. The chief reason is the lack of local rock material suitable for producing the aggregate needed for a bituminous asphalt surface.)

In the Newcastle section, design engineers used the topography to provide a pronounced difference in the elevation between the eastbound and westbound lanes. West of the Ganaraska River, for almost a mile downgrade the median widens from 34 feet at the top of the hill to a maximum of 125 feet before it begins to taper to 34 feet where the four lanes again reach the same level.

Although diverse driving features are desirable (they relieve the monotony of straight roads and thus keep drivers more alert), it is economically feasible to incorporate them only under certain conditions. Hilly topography allowed six such sections, totalling five miles, to be included in the 28 miles between Port Hope east to Brighton, and again, on sections between Brighton and Trenton.

Safety features

For the earliest sections of the freeway the median between the lanes was 30 feet. Later the width was increased, first to 34 and then to 36 feet. Still later the 54-foot medians appeared and for quite a period, the width selected for additional sections varied between 50 and 60 feet. After that came sections with medians of 70 feet, or more.

Safety was the prime reason for these increases. The wider the median, the more room for manoeuvring in case of emergency and the less danger of crossing over to the opposite lanes. Similarly, the problem of headlight glare fades as median width increases.

In addition, it is easier and cheaper to add extra lanes to sections where space already exists, and the original bridges and overpasses are already wide enough to contain them.

The designs of interchanges and bridges were usually modified for each new section, with planners continually looking for the right combination of aesthetics and economy. The cloverleaf design, trademark of the earlier sections, was altered greatly in later structures, and interchanges became more sophisticated. In the 30 units which form the trans-provincial route, it is fairly easy to spot some of these evolutionary stages.

On some of the earliest portions there are one- and two-span sections, then, as the median was widened, the designers switched to three- and four-span structures. (A span is the horizontal section carried between vertical supports. These supports can take the form of piers, columns or the abutments—the resting shelves—at either end of the bridge.)

Four-span bridges feature open abutments, so described because of the open space or "light" between the underside of the bridge deck and the bridge embankment sloping down to the shoulder of the roadway. As a safety measure, this design allows aproaching traffic on the freeway to see through the open spaces at each side of the bridge and to spot vehicles entering or leaving interchanges.

The gap is closed

Finally the day arrived when the entire highway was opened from Windsor to the Quebec border—even though some of it was still two-lane sections. The official opening was held November 10, 1964.

The final gap was a four-lane stretch of slightly less than 10 miles east of Lancaster. Two years later, the Quebec Roads Department continued the four-lane connection to Montreal with its controlled-access Route 20.

Another memorable date in freeway history was October 11, 1968, the day a new 15-mile section from Gananoque to west of Brockville was opened. It meant there was now a minimum of four lanes over the full 510-mile route.

Department of Highways, Ontario

Plaque unveiled on October 20, 1969, by The Honourable James Auld, Ontario Minister of Department of Tourism and Information, on left, and The Honourable George E. Gomme, Ontario Minister of Highways, to mark the completion of 401.

This final mileage—which included an interchange at Highway 137 connecting the freeway directly with the Ivy Lea Bridge, The Thousand Islands and the United States border—replaced a 25-mile section between Gananoque and Long Beach which had served as the main route for the previous 25 years.

LOOKING TO THE FUTURE

What is the future of the 401 and other freeways? Some believe that by the year 2000 there will be a "city" stretching from Windsor to Montreal. Instead of just one main route, there will be a series of superhighways running the 600 miles.

Certainly the start is there for such a megalopolis. The heavy concentrations of industry reaching out from Toronto and starting to build from other major centres along the route could eventually unite into one major concentration.

But this vast build-up of people and industry will also surely bring changes in modes of transportation. The automobile of today, with its gasoline-powered engine, may be modified or ruled out in the future because of new air pollution regulations and controls or the introduction of new power sources (such as electrical or solar heat storage batteries or nuclear energy). Entirely new concepts in travel could be introduced.

Already designers talk of vehicles zipping along on air cushions at 300 miles an hour, computer-controlled and radio-dispatched cars and buses, or commuter trains floating on compressed air through subway tubes. These and other means of urban transportation are the visions of tomorrow.

The motorist of the future, ready to start on a trip, may well consult a destination code book (which could resemble a postal zip-code), find the code word for his destination, insert it in an electronic unit in his car, and set out on the freeway with never a care about roadside signs, maps or any other source of information.

As his vehicle approached each intersection, the destination code would be transmitted to roadside equipment and then decoded in accordance with a stored programme. An appropriate manoeuvre instruction would be calculated, returned to the vehicle and visually displayed to the driver. Instructions, such as "turn left" or "turn right" and directional arrows would appear on the dash or windshield of the vehicle. The entire information exchange would take place in the length of time necessary for the vehicle to travel a distance of five feet at highway speeds.

This type of computer-age motoring may sound distant, but already the development of this experimental route guidance system is under study by the United States Bureau of Public Roads, and by private industry.

Also under study is a "dial-a-bus" idea for resolving future traffic congestion. Instead of each commuter operating his own car, radio-dispatched vehicles would pick up transit users at their doorsteps and deliver them to their destinations.

The thought of congestion in a country the size of Canada is often hard to comprehend. But the fact remains that in 1970 roughly half of all Canadians lived in metropolitan areas with populations of over 100,000: a one-third increase since 1951. By 1980 it is estimated that six out of 10 Canadians will reside in urban areas with metropolitan populations of over 100,000. By then, one-third of the country's population will be living in Montreal, Toronto or Vancouver.

Traffic problems will be enormously increased. The total number of passenger cars in Canada reached a million in 1929, a figure which remained almost constant until the late 1940's because of the depression and the war years. But then registration began a rapid climb, doubling by 1952 and doubling again in the next decade.

With the new cars came demands for new roads and the 401 is just one of the many major expressways which have been built across the land in the post-war years. In 1945 there were less than 24,000 miles of surfaced roads in Canada. There was a 100% increase by 1955 and by 1968 the figure had reached 106,000. Of this last total, more than 1,500 miles were of freeway standard.

During the same post-war period, federal and provincial co-operation created the Trans-Canada Highway. This roadway, numbered differently in each province, runs the entire length of the country, from St. John's, Newfoundland, to Victoria, British Columbia, and has been constructed to a specified standard.

Economic Council forecast

The Economic Council of Canada, in its sixth annual review issued in 1969, says the Canadian transportation scene has been marked by three prominent features over the last decade or so. In the urban areas, there has been a marked decline in the use of public transportation facilities which has resulted in a corresponding increase in the use of the private automobile. Intercity movement of both goods and passengers has been rising more rapidly than national output, with increasing shares of the total carried by airlines and pipelines. At the same time public expenditures on transportation have grown more slowly than other major areas of government spending such as health and education.

The Council stated:

"Our estimates suggest that the growth in transportation expenditures should become somewhat more rapid than it has been in the 1960's though still less than the growth in health and education. By

1975, governmental spending on transportation may exceed $4 billion (in 1967 prices).

"Spending on highways, roads and bridges, which now takes up four-fifths of transportation expenditures, is expected to show a rapid rate of growth with the largest increase projected for urban areas. With the completion of the Trans-Canada Highway program, the number of large-scale highway projects contemplated is considerably reduced. However, urban requirements, particularly arterial roads and bridges in a number of cities, remain very pressing.

"In Canada, construction of roads and streets now absorbs 3 per cent of national output. Some increase in this proportion is anticipated to accommodate the increasing volume of road traffic—both intercity and urban—the increasing demands for access to recreation areas arising from higher incomes, and the probable requirements for multilane limited-access highways to link major urban centres. However, our projections are conservative; the backlog of urban transportation projects may well continue to rise against a background of increased traffic congestion in many locations. In these circumstances it is important to develop better planning of urban transportation systems and to acquire appropriate rights of way to facilitate the improvement of mass urban transit facilities."

Vehicles and air pollution

With this problem of congestion comes another: air pollution. The situation is more dangerous in direct proportion to the volume of cars and trucks operating in built-up areas.

Although air pollution and its attendant health hazards are by no means entirely created by motor vehicles, cars and trucks do contribute a considerable share of pollutants to the atmosphere. For instance, Toronto City Council Alderman Tony O'Donohue made this rather startling statement in 1970 during an appearance before a public inquiry held by Pollution Probe and GASP (Group Action to Stop Pollution), two organizations of academics, students and ratepayers:

"The total output in Toronto every year is 950,000 tons of carbon monoxide and about 97% of that comes from the automobile. Another problem is hydrocarbons, of which we have about 500,000 tons a year. About 85% of the hydrocarbons come from the transportation industry."

The problem—to clean up the internal combustion engine—is a very difficult one, namely the reduction of exhaust emission carbon monoxide and hydrocarbon pollutants. Carbon monoxide is highly toxic and so are some of the hydrocarbons. The two are components of smog.

A characteristic of the gasoline engine is that the output of smog ingredients diminishes at higher speeds (although it then reappears as a problem at very high speeds). The condition is worse in slow speed, stop-start traffic where engine combustion is seldom complete and partially burned fuel passes the exhaust system as smog waste with high pollutant levels.

Thus, larger engines which tend to "loaf" most of the time when their full power cannot be utilized are more difficult to control than smaller engines working much nearer their maximum output. Heat is critically important in the suppression of exhaust emission. Unfortunately the amount required for uniform emission control is difficult to provide under all driving conditions.

The transportation industry is not the only offender. Alderman O'Donohue also said that about 150,000 tons of sulphur dioxide enter the atmosphere each year in Toronto, 70 to 75% from power plants.

"The organics emitted into the air in the form of pollution are about 220,000 tons a year, and we have about 40,000 to 50,000 tons of inorganics and solids. So we have quite a significant problem here", he stated.

A Toronto radio announcer, Eddie Luther, testifying at the same Pollution Probe and GASP inquiry, described what this combination of smoke, dust and other pollutants looked like from the high-level view provided by a traffic helicopter.

"Many mornings you can actually taste it; on some mornings it forms a very heavy blanket over the city . . . just the other day we had clear skies, a nice blue sky. We climbed to about 2,000 feet above the ground and at that point we were just at the level of this blanket, a flat blanket. Below that it was dirty brown, smelly— some days it will cover just a small area downtown. Some days it will cover the east end or the west end, or it may be miles away from the city. . . . Most of the time it covers basically the downtown part of the city."

Although this was the smoggy scene presented for Toronto, the same grimy picture could be painted for almost any major city on the North American continent.

There are other factors in air pollution, but what can be done about the problem created by cars and trucks? There is certainly recognition that the internal combustion engine and its exhaust "emission" must be controlled. This is evident by the legislation adopted, or in the process of adoption, by several Canadian provinces. Ontario has had regulations of this kind since 1968, while Quebec and British Columbia are studying programmes of their own.

Montreal in 1970 passed a bylaw which prohibited drivers of parked cars from idling their engines for more than four minutes. Up to 60 days in prison or a maximum of $100 in fines were prescribed for offenders.

But while the legislation is Canadian, most standards are American, since many of the vehicles sold in Canada are now built in the United States under Auto Trade Pact provisions. So whether the customer lives in a province with automotive pollution controls or not, his new car has them anyway (at an added cost of $35.00 to $50.00).

The United States Government has had a federal automotive emission code for several years (which individual states may adopt or even enlarge upon), while California introduced its own specific code in 1965. In fact, much of the original California code was used as the basis for United States federal and subsequent Canadian regulation.

Generally speaking, present legislation is progressive to the extent that the amount of emission permitted is diminished annually. For instance, 1970 models produced approximately 70% less emission than pre-emission control vehicles. The objective is a 90 to 95% reduction by the mid-1970s. Many authorities and engineers feel the internal combustion engine will be outmoded in the near future with alternative power plants offering inherently clean exhaust, typically the gas turbine. This type of plant has already proven successful in trucks, buses and heavy vehicles.

The code covering the amount of emission an engine may produce is a complex one based on engine size and output variables. Each engine is tested and certified (in the U.S. by a federal agency which supplies this data to Canadian authorities) in conformity with the code.

A new emission problem is emerging as a result of hydrocarbon research. This is the increasing significance of nitrous oxides, which some authorities believe result from the lead content in gasoline.

Now that some of the causes of exhaust emission have been identified, urban planners and highway engineers are attempting to speed up the flow of traffic with new types of feeder lanes and crossovers leading into congested areas. The higher speed improves traffic flow and, through more efficient engine operation, minimizes output of pollutants.

Private car or mass transit?

In the 1970's, almost every level of government is facing major transportation problems. It is both a question of finding the financial means of providing improved travel facilities—in the face of demands for other social services—and also in deciding what form the new travel facilities will take.

In the major cities there is an increasing emphasis on rapid transit routes, such as commuter trains and subways. But high-speed express-ways are also being added to cope with rising car ownership. And there will likely always be the commuter who wants the comfort of door-to-door travel in the convenience and privacy of his own vehicle. However, catering to this demand is expensive. A mile of elevated expressway in an urban area can cost up to $16 million, with 60% of money going just for the acquisition of the land. In the future, the private car may be banned from some downtown areas to avoid total traffic chaos, and possible destruction of the character of the city. Some planners and sociologists believe that freeways can bring death rather than life to a city.

Where a city core just cannot absorb any more vehicles, an alter-native is to develop rapid transit systems whereby commuters are encouraged to drive to a station and take a train to work. Proponents of subways and train travel point out that to carry in automobiles the 30,000 people hourly who can be moved in reasonable comfort by a rapid transit line requires 15 freeway lanes and 30 arterial lanes at a substantially higher cost. In addition, depending on load factors, a car travelling at 20 miles per hour uses 6 to 45 times the land space per person used by a transit bus.

But the competition between the two modes of travel, both for financial backing and passenger support, is likely to continue into the foreseeable future, with a judicious blending of both seemingly the most likely answer to the more immediate demands.

What is required, however, is more integrated planning, with aggres-sive stands taken to design and ensure there is more liaison and harmony in the development of related forms of transportation.

CHAPTER 2

A City and its Heart

AN IDEA GROWS

The construction industry, more than almost any other, profoundly affects the lives of people. It is both a determinant and an indicator of the direction of the economy. It touches on business, commerce, transportation, manufacturing, worship, the home, recreation and entertainment. It changes behavioural patterns of a village or a nation.

Since 1956, downtown Montreal has been one huge construction project valued at more than a billion dollars. Many cities have tall buildings and some, like New York, have taller ones than Montreal and more of them. But in exploitation of the principle of separation of forms of movement, Montreal has become a world leader.

This separation of types of traffic involved development of a subsurface city consisting of passageways and corridors located under a series of magnificent new buildings, all erected in the 1956-1966 period. These complexes were the skyscrapers Place Ville Marie, Place Victoria and Chateau Champlain (part of Place du Canada), and Place Bonaventure, a structure of moderate height but just as impressive in its own way. Important new buildings that do not form part of the subsurface city are the Hydro-Quebec building, the Canadian Imperial Bank of Commerce Building and the Canadian Industries Limited Building, known as CIL House.

The development of downtown Montreal provided a stimulus and an opportunity for achievement to the men and women of varying backgrounds who worked on the projects. But it was to affect in greater or lesser degree all the two million and more persons making up the city's population.

Planners are fond of saying the principle of traffic separation was first enunciated by Leonardo da Vinci in 1495. Da Vinci prepared sketches for a city of his era showing pedestrian levels above traffic with walkways spanning the streets and arcades penetrating buildings. The world generally has been slow to adopt the principle but in 1913 an Italian architect, Antonio Sant' Elia, prepared a sketch showing trains, trucks, cars and people all occupying their own private levels within a framework of tunnels, slim-line bridges and towering sky-scrapers.

There are many examples of the principle in action, although mostly on a modest scale. The simplest form is separation by time, that is, an area is reserved for the sole use of pedestrians during certain hours. This method is used in some European cities. The form most often seen in Canada is horizontal separation. This is the method employed at such large shopping centres as Yorkdale in Toronto. Here a system of covered streets is set in the middle of a huge surrounding parking lot. The car is left outside while in the middle the pedestrian has the run of the streets. The arrangement is similar at Ottawa's Sparks Street mall and the Water Street mall in St. John's, Newfoundland. In each case, a few blocks of the main shopping area have been closed to vehicular traffic.

Vertical separation, the truest form, employs a variety of techniques and is the method found in Montreal. Using this system, pedestrians, rapid transit, trains, trucks and cars all have their own levels and move about in them freely, each unhindered by the movements of the others. The principle involves taking advantage of topographical features to put pedestrians at many levels, sometimes below the vehicles in passage-ways and sometimes above them in elevated walkways and bridges.

There has been occasional mild criticism of Montreal's system on the basis that the multi-level pedestrian approach has not been fully devel-oped. British architect Norbert Shoenauer has noted that the 45-foot escarpment adjacent to the Place Ville Marie site provides the perfect opportunity for enclosed pedestrian bridges connected to the complex. (A small open pedestrian bridge now links the Place du Canada plaza with Dominion Square across Lagauchetiere Street.)

Another such system is New York's Rockefeller Centre, with 19 buildings linked by a series of austere tunnels winding through 17 acres. This network was started in the 1930's and its unyielding atmosphere makes pedestrians glad to be back on traffic-crowded streets. But lessons have been learned from this not very successful start and better results have been achieved in Philadelphia, Pennsylvania, in Boston, Massachusetts, where about six acres have been developed and in Hart-ford, Connecticut, with a plan covering four acres.

Montreal's system does provide separate surfaces for all forms of movement. In its streets and malls, the pedestrian is permitted complete forgetfulness of foul weather and traffic hazards. Two lower levels provide parking space and trucking facilities that can be reached only from the edges of the system or as far away from Place Ville Marie as four blocks. The Metro subway has its own private route and the whole for the most part is underlain by 16 sets of railway tracks.

Vehicular traffic has always presented a problem for Montreal, which is faced with the need to squeeze large volumes of it between two of the city's important geographical features, Mount Royal to the north and the St. Lawrence River to the south.

Philadelphia's multilevel development is probably the most ambitious in North America outside of Montreal. About 165 acres are to be developed but a relatively small part of this programme has been carried out. Subway passages linking the main parts of the system are dotted by benches and shrubs and natural light streams through openings in the ceilings. Montreal development was undertaken entirely by private capital while in Philadelphia, prime initiative came from the city.

Vincent Ponte, the private town planning consultant who worked with the owners and architects in the design of both Place Ville Marie and Place Bonaventure, has suggested introduction of municipal bylaws in major cities where the possibility of such projects is foreseen. These would require the owners of new buildings to provide spaces above or below ground that could be plugged into the pedestrian grid as it approaches.

Ponte, who travels the continent working on such projects, told a group of zoning officials in Dallas, Texas in 1968 that zoning regulations, like cities themselves, have grown haphazardly in a pattern dictated by a press of isolated local problems and that perhaps it would be a good thing to rationalize them with a long-range view to the welfare of the city as a whole.

He noted that zoning in downtown areas today is chiefly concerned with controlling land use, density, and providing sufficient light and air for the streets. Thus the effect of zoning is vertical while the multilevel city concept is concerned with horizontals, with decks and levels extending blockwise across the core, over and under existing streets.

Ordinances requiring developers to create horizontal facilities that could be connected to the system as it expands would benefit both the owners of the buildings and the public at large, says Ponte. The space could be used by the owner in any way he sees fit until the time comes to link it to the system.

Canadian Pacific

Trizec Corp.

Place Victoria-St. Jacques Co. Inc.

ffleck, Desbarats, Dimakopoulos, Lebensold Sise, Montreal

Facing page:
Place du Canada complex
Above left: Place Ville Marie
Above: Place Victoria
Left: Place Bonaventure

The idea of using one building for many different purposes is not new. But the trend is accelerating and now almost any large city has at least a few examples. Such multi-use buildings are Place Ville Marie, Place Bonaventure, Place Victoria and Place du Canada.

The most complete and unusual multi-use building embraced by Montreal's system is Place Bonaventure, since it incorporates retailing facilities, transportation services, office space, an international merchandising centre, a huge trade fair floor and a 400-room roof-garden hotel.

Place Ville Marie is multi-use in that it provides retail-commercial facilities, office rental space and through Place Bonaventure, access to the Metro. Place Victoria offers retail and office space with Metro access and Place du Canada offers hotel accommodation, office space, retail facilities and quick access to both the Metro and trains. All of these buildings provide their own parking and service garages.

Critics have noted that the multi-use character of these buildings represents a desirable break from the inflexible single-use zoning patterns that have characterized many city centres in the past. But some say the break with tradition has been only partially accomplished because of the lack of residential accommodation aside from hotels.

The planners involved in the Montreal projects say there is no possible way residential use can be introduced into these developments because it would be so expensive that not even millionaires could afford it. A constant income, a flow of money far in excess of the area's real residential value, would be required for tenancy. A better idea, the planners say, would be to initiate high-density residential development around the periphery of the core, possibly along Sherbrooke Street, between Stanley Street and Beaver Hall Hill.

If the pedestrian grid system reaches Sherbrooke Street by the late 1970's as planned, this type of residential development would bring thousands of persons close enough to walk to work each day and to visit it often for after-hours recreation or entertainment.

Residential accommodation has found its way into some improbable use combinations in recent years but usually in locations away from the core of the city. In New York, where the multi-use trend itself has grown faster than elsewhere, some buildings include apartments, shopping facilities, theatres and schools. Here a young student could live, entertain himself and get an education without ever going outdoors.

What one critic calls the "purest" residential-commercial complex in Montreal is the Westmount Square development outside the core but just inside the western boundary of the central business district. This development consists of an office building with a shopping centre at its base and two apartment buildings. In Chicago, the 100-storey John Hancock Building is almost equally divided between apartments and offices. But here again, the building is located outside the core of the city.

Another example of this type of development is the Colonnade Building on Bloor Street in Toronto, which incorporates apartments, offices and shopping facilities with restaurants and a theatre. The Toronto-Dominion Centre and the Richmond-Adelaide Centre, both in Toronto, combine office and retail uses.

A concept found in Montreal, and one that planners feel is necessary for the best results in downtown renewal, is that of the superblock. A superblock is a block of land developed as an entity rather than on a piece-by-piece basis. Thus, instead of 20 small buildings in a single block, each containing offices, a superblock is one building, or perhaps more as long as they are designed to form an integrated unit, capable of meeting the space requirements of many tenants.

The essence of the superblock is organization. A building is designed for compatibility with its surroundings or a series of buildings are designed to be compatible with each other and their surroundings. And in either case, the design complements the city itself and makes the best use of topographical features.

This type of organization invariably creates elbow room, traffic-free open areas to be enjoyed by people. Concentrations of people make introduction of secondary uses more feasible. The whole offers an opportunity for the exercise of total urban design. These super-blocks can be linked in the manner discussed as being characteristic of the multi-level city. In Montreal, Place Ville Marie, Place Bonaventure, Place Victoria and Place du Canada are examples of super-block development.

CITIES ARE FOR PEOPLE

"If I couldn't work here, I would probably leave the country", is how a legal secretary spoke of Montreal's Place Ville Marie. "I think it is marvellous. People who work here can get everything they need without even going outside the building." "Everything" includes dresses, shoes, groceries, books, records, cocktails, lunch and dinner, jewellery, sporting goods, movies and a theatre!

This enthusiastic secretary is experiencing multi-level living in one of North America's most exciting city centres consisting of four ambitiously-conceived complexes created during the "Golden Decade" of 1958-1967. Her views are echoed by many of her 22,000 counterparts in Place Ville Marie, and by many more in Place Bonaventure, Place du Canada and Place Victoria.

Commented a young secretary in Place Victoria: "I never buy anywhere else unless I can't find what I want in the complex and that seldom happens." She considers the selection of goods and prices to be as competitive as she would find closer to her home. She lives in west-end Montreal and a 25-minute bus ride takes her to work. She even buys some of her groceries in the big grocery store in the complex and carries them home on the bus.

Still another woman, also in Place Victoria, stresses the convenience and sociability of the complex; the girls visit each other during lunch hours or tour the shops as a group. She and her husband live with her mother, who looks after the grocery shopping, but she buys everything else she needs right in the complex. And she notes that if a change is wanted from the 40 stores in Place Victoria, there are 100 more across the street in Place Bonaventure.

"Golden Decade" of construction: 1956-1965

Place Ville Marie, built at a cost of about $110 million, was opened in 1962, Place Victoria, which cost $90 million, in 1965 and the $88-million Place Bonaventure in 1967, just in time for the Expo 67 trade. About the same time, and also aimed at gathering in some of the Expo trade, Canadian Pacific's $40 million Place du Canada complex was opened. This complex consists of a 640-room luxury hotel and a 27-storey office building, both rising from a plaza over a shopping arcade.

Of the years making up the golden decade, 1962 must surely have been one of the brightest. In the spring of that year, three other major buildings were opened within a span of two months. Diagonally across the Dorchester Boulevard-University Street intersection from Place Ville Marie, Canadian Industries Limited opened a $30-million, 34-storey head office building sheathed in an image-reflecting combination of glass and black porcelain. Two blocks west of Place Ville Marie, the Canadian Imperial Bank of Commerce opened a sheer-rising, 47-storey tower of green-grey slate, glass and steel. Eight blocks to the east, actually far enough away to be just outside the city core, Hydro-Quebec opened a $23-million 26-storey headquarters building of sand-coloured stone.

Of the four projects opened in 1962, Place Ville Marie was the first under construction and the last to be completed. Planning for

Place Ville Marie began in 1956 and so was indisputably the spark that ignited the boom. Construction started in 1958 and the complex, with only two of the ultimate four buildings completed, was opened in September, 1962. The three other complexes were started in 1959 and opened to the public in the spring of 1962.

But the hub of this bright new downtown is Place Ville Marie, with its broad three-acre plaza in a pattern of deep-pink aggregate squares separated by concrete, its unique 47-storey cruciform tower and three smaller buildings ranged along two sides of the square bounded by Dorchester Boulevard, University, Mansfield and Cathcart Streets.

The pedestrian ways

Beneath Place Ville Marie, Place Bonaventure, Place du Canada and Place Victoria, sometimes following original grade levels and sometimes underground, are covered, climate-controlled pedestrian walkways with glass-walled or open-front bazaar-like stores and shops, cinemas, restaurants and service establishments.

Place Ville Marie, Place Bonaventure and Place du Canada, along with the older Central Station and Queen Elizabeth Hotel, are connected by this system of corridors, which makes frequent use of the mezzanines and passages of the city's subway system. Many of the corridors themselves are shop-lined, making them integral components of the pedestrian system.

Thus it is possible to walk four city blocks, almost in a straight line, from St. Antoine Street to Cathcart Street without exposing oneself to the weather. The route lies through Place Bonaventure, the Canadian National headquarters building, the Queen Elizabeth Hotel and Place Ville Marie. In fact, the back door of Place Ville Marie, if such a many-portalled complex can properly be said to have a back door, opens up aisle-like on McGill College Avenue, a mere 200 feet from St. Catherine Street, the main commercial and shopping artery of the city. And through the Place Bonaventure Metro station, one may walk under cover two additional blocks to Canadian Pacific's Windsor Station.

This three-mile-plus network of covered streets and malls provides three million square feet of retail area including 200 shops, 35 restaurants, several movie houses and a live theatre. Within the overall framework of the system are 2,500 hotel rooms in three hotels, parking spaces for 8,000 cars, two subway stops and two railway stations. Four great office buildings provide 13 million square feet of prime office space and there is an international trade centre.

PEDESTRIAN PASSAGEWAYS

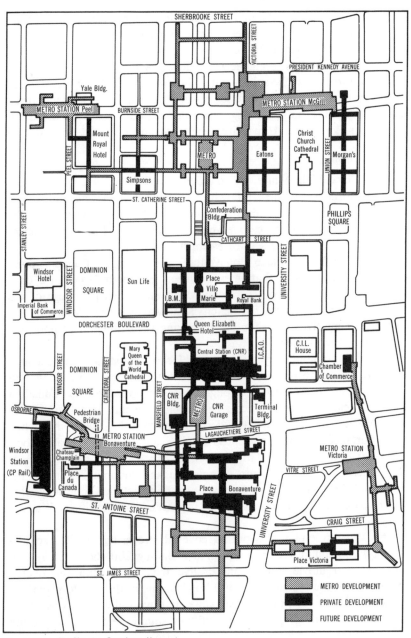

Vincent Ponte, City Planning Consultant, Montreal

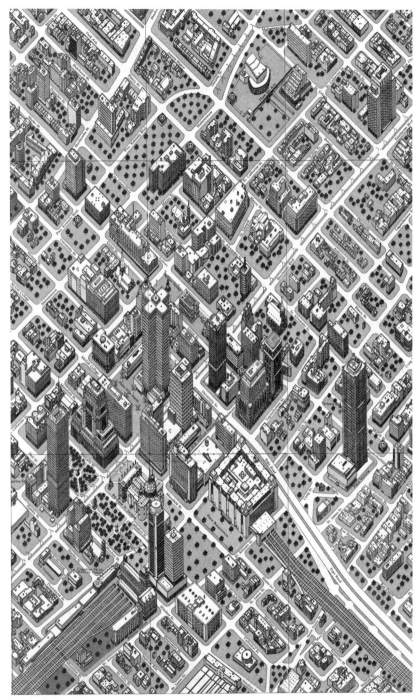

Penthouse Studios Inc., Montreal

Threaded through about 70 acres of the city core, the total complex is capable of accommodating 400,000 persons none of whom need cross a street intersection at any time during the day or get his nose nipped by frost as he pursues business or pleasure.

And the end to development of this controlled-environment system is still years away, if indeed there is an end. By 1970 the impetus of 1956 seemed to have spent itself for the time being but at that time plans were firm for further development to the immediate south and west of Place Bonaventure and for an additional twin tower on the Place Victoria site, all of which are to be plugged into the system.

The development area for which plans are already in existence embraces 185 acres bounded by Stanley, Sherbrooke and Notre Dame Streets and Beaver Hall Hill.

The Eaton's, Morgan's and Simpson's stores, all within a few blocks of each other on St. Catherine Street, already have underground shopping areas which will eventually be linked up with the system by a corridor from Place Ville Marie under McGill College Avenue. The development will then continue north past Burnside Street and President Kennedy Avenue to Sherbrooke Street and the gates of McGill University, nestling at the foot of Mount Royal.

Surface plans in this area include a widened McGill College Avenue with corridor-forming buildings of uniform height arcaded at street level and designed to make the street, as the planners put it, "more than just a long hyphen" between the Mountain and Place Ville Marie. One will have an almost unobstructed view of the mountain from Place Ville Marie or from any point on McGill College Avenue.

Plans for this development of the northern section of the downtown core have undergone considerable change since inauguration of the system and will probably evolve further, but these are the basic proposals. Westward, the system already extends to Windsor Station from Place Bonaventure and later will probably move westward again on St. Catherine Street to take in the Mount Royal Hotel at Peel Street.

CIL house, the Bank of Commerce Building and Hydro-Quebec are not connected to the pedestrian system. Hydro-Quebec, a public utility, is somewhat distant for convenient linkage and none of these buildings incorporates shops or entertainment facilities. Below CIL House is a restaurant and bar.

After this brief introduction to the buildings, we are ready for a tour of the pedestrian network by the longest route. In any such tour, we cannot afford to bypass Place Victoria, even though it is not yet integrated with the system.

Place Victoria is on historic St. James Street, the birthplace of Montreal and since the mid-1800's, the financial and trading centre of the city. St. Catherine Street, and Dominion and Philips Squares with their high concentration of shops, department stores and restaurants, form the commercial core.

The home of many banks, national corporations, trust firms and the Montreal and Canadian Stock Exchanges, St. James Street has declined in importance and influence in recent years while St. Catherine Street has prospered. And now the prosperity of St. Catherine Street has over-flowed to Dorchester Boulevard. Thus Place Victoria's location suggests its builders wanted to show their faith in that part of the city.

The Place Victoria shopping Arcade is similar in decor and atmos-phere to that of Place Ville Marie. The shops are glass-fronted and the lighting is of a subdued brightness. There are 35 stores, shops and art galleries, five restaurants and a 530-seat cinema. Retail establishments are arranged in a unique manner in that the so-called fashions are separated from service-type stores. The upper level offers clothing, jewel-lery and sporting goods while the lower level, from which one may also board the Metro, has a grocery store, post office, hairdressing services and restaurants. The purpose of separating types of businesses, says the management, is to reduce traffic and to encourage among consumers more systematic shopping.

In a long, five-storey building adjacent to the tower are located the stock exchanges, formerly housed in an earlier building on the same site. The room occupied by the exchanges is rich-looking with plush-red carpeting and a special ornamental acoustical system of white asbestos characters on upper walls and ceiling design to deaden the sound of ticker-tape machines and the calls of buyers and sellers. Visitors may watch the market in action from a gallery along one side of the room.

To tour the main system, we start at the entrance to Windsor Station on Osborne Street. Under the canopy of Windsor Station, we descend a flight of stone steps to the Metro level and follow right-curving corri-dors to the Bonaventure subway station. As we enter the subway sta-tion, we can turn sharply right to find an entrance to the Place du Canada complex, the 640-room Chateau Champlain and its accompany-ing 27-storey office building. Beneath the office building is a shopping concourse with restaurant and cinema.

Continuing our walk through the station, we cross the subway mezza-nine from which we may pause to watch the blue-painted rubber-tired trains arriving and departing beneath our feet. Ascending an escalator, we find ourselves in a boutique- and shop-lined avenue. This is the lower level of the Place Bonaventure shopping concourse.

Along the corridor is a central court area from which another escalator carries us to the upper level. Here the store fronts seem to step out to beckon us in. The secret is in the lighting technique. The "street" lighting, as construction men call the lighting of corridors and large open spaces, is deliberately kept dim while the shop interiors are illuminated fairly strongly, causing them to stand out. Enhancing the overall effect are the vivid colours merchants use in dressing their store fronts. One of the sharpest of these contrasts between Place Bonaventure and Place Ville Marie is in the approach to illumination.

Above the shopping concourse is the five-acre Concordia Exhibition Hall, with its great tree-like columns supporting the tons of surrounding concrete. And above that are five floors of offices and a wholesale merchandising area.

Perched atop Place Bonaventure is the Bonaventure Hotel, with its roof garden of trees, shrubs, waterfalls, streams, swimming pools, indoor and outdoor restaurants and park benches. The 402 rooms and suites are banked in four storeys along the periphery of the roof, while restaurants and administrative offices are grouped in the centre. The garden wanders for three acres between them.

Place Bonaventure, perhaps more than any other of the components of the system, offers attractions for younger members of the family who may have no particular interest in shopping. Musical groups and singers are often featured and Concordia Hall is used for entertainments and exhibitions of interest to small children and teenagers, as well as for big trade shows.

In 1970, there were sections of Place Bonaventure still unfinished and the building was not being used to capacity. Officials noted that the major service Place Bonaventure provided, a forum for the exhibiting and promotion of merchandise for international buyers and sellers, was new to Montreal and that exhibitors had been in the habit of displaying in such centres as Chicago.

One can easily become lost in either Place Bonaventure or Place Ville Marie the first time out, probably because neither are perfectly uniform in their corridor layouts. There are three east-west "streets" on the upper level of the Place Bonaventure shopping concourse and only one running the length of the building in a north-south direction. The Place Ville Marie shopping promenade is almost rectangular in shape. But if one wishes to make a complete circuit, the rectangle is interrupted at the east end by escalators and a short walk through the Royal Bank Building.

To leave Place Bonaventure, we walk through an underground corridor called The Bonaventure Walkway to the Canadian National headquarters building.

Affleck, Desbarats, Dimakopoulos, Lebensold Sise, Montreal

This is Place Bonaventure's upper shopping concourse. The escalator leads to a second one, which gives access to the subway system.

From the CN building, a broad corridor leads into the south-west corner of Central Station. From here, there are several alternatives for reaching the Queen Elizabeth Hotel, the heart of the system. The direct route leads across the station and through the lower lobby of the hotel. Alternatively we could walk eastward the length of the station and through a long, somewhat austere corridor directly into the Place Ville Marie shopping promenade.

As we step into Place Ville Marie, the lighting contrast with Place Bonaventure is obvious. The lighting still has a subdued quality but the overall impression is one of brightness and gaiety. The impact of individual façades on the observer is slightly reduced and the atmosphere is one of less formality and perhaps reduced opulence.

The Place Ville Marie type of development, while not completely untried (Rockfeller Centre was an earlier attempt), was rare in 1962 and the developers accepted almost any business seeking space. No attempt was made to sort stores according to price, type of merchandise or other criteria. But success came quickly and the owners are now in a position to pick and choose. A form of order has evolved best described by one official who says "there is an area for big stores and an area for small stores." One result of this evolvement has been a "street" known as Fashion Row.

During the tour are passed dress shops, a book shop or two, men's wear stores, a sports shop, shoe stores, candy shops, clothing stores catering to all tastes. There are antique shops, a furniture store, a rare coin shop, several knitwear and woolcraft shops, camera shops, leatherwear shops and several restaurants. There are stores specializing in Canadian handicrafts and Eskimo art.

Within the system one may book a hotel room, arrange train or plane passage to Vancouver or Florida, arrange a consumer loan, attend to family banking matters, get a quick snack or a full course meal, take in a movie or a play.

We have reached Fashion Row. To our right, as we turn the corner, are floor-to-ceiling windows through which is visible one of the four court areas providing access to the promenade from the plaza above. Standing close to the windows, we look up at the huge cruciform tower. Framed by the concrete rail of the stairwell, the tower's great height is even more impressive from our recessed vantage point than from plaza level.

At the end of the Fashion Row an escalator to our right leads to the main lobby of the Royal Bank tower. To our left is a street leading to the cinemas. There are two movie houses, a large one offering standard movie fare and a small one featuring exotic or art films from abroad. Occasionally, when a film has had a good run in the large cinema, it will be continued in the small one for the benefit of stragglers.

We ascend the escalator to the lobby of the Royal Bank building. The lobby here is 57 feet high, the equivalent of five stories. A short flight of steps and another escalator ride would take us to the main banking hall of the Bank, at 300 feet long one of the largest banking halls in the world. The hall is located in the two western quadrants at the base of the tower, thus making possible natural light through a series of translucent domes in the ceiling.

After a brief walk through the lobby, we descend by an escalator to the shopping promenade again. Another escalator goes down to a junction with the austere, shopless corridor leading from Central Station

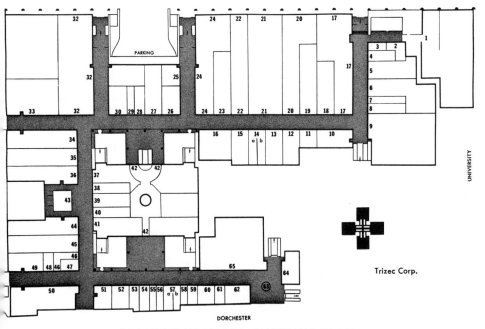

CATHCART

PARKING

UNIVERSITY

Trizec Corp.

DORCHESTER

PLACE VILLE MARIE: SHOPPING PLAN

1 Theatre	22 Ladies' wear	45 Toys
2 Art gallery	23 Shoes	46 Stationery
3 Candy	24 Pharmacy	47 Candy
4 Bookstore	25 Duty free shop	48 Rare Coins
5 Stationery	26 Boutique	49 Boutique
6 Clothing	27 Copper shop	50 Liquor Board
7 Jewellery	28 Millinery	51 Bookstore
8 Hosiery	29 Jewellery	52 Photographic centre
9 Cafeteria	30 Shoes	53 Stereos and TV
10 Cigars	32 Ladies' wear	54 Linen
11 Children's clothing	33 Supermarket	55 Football club office
12 Boutique	34 Bridal	56 Real estate
13 Bakery	35 Barber	57a Leather
14a Knitwear	36 Men's wear	57b Wool
14b Boutique	37 Ski shop	58 Hotel reservations office
15 Shoes	38 Gift shop	59 Florists
16 Cafe	39 Records, Bookstore	60 Cleaners
17 Ladies' wear	40 Oriental	61 Optometrist
18 Men's wear	41 Jewellery	62 Bakery
19 Shoes	42 Four restaurants	63 Cigars
20 Ladies' wear	43 Travel Agent	64 Bar
21 Ladies' wear	44 Boutique	65 Trust company

that we noted earlier. At this junction, in a depression at the southwest corner of the promenade, is located the 99-seat "instant theatre" which at noon hour presents three performances of half-hour plays, often by such well-known playwrights as George Bernard Shaw and Edward Albee.

After buying tickets, one can pass the sandwich and coffee counter to select lunch to enjoy with the entertainment. The house is almost invariably sold out for each performance. Here we can see brief-cased businessmen, stenographers and sales girls, visitors to Montreal, and housewives seeking respite for weary feet.

Studies of pedestrian traffic through Place Ville Marie indicate there are almost 200,000 individual comings and goings through its 17 entrances every day. Many persons will be counted twice or more, but the head count is the best method yet found of determining usage of the facilities. No similar counts are available for Place Victoria or Place Bonaventure but tours of the three complexes at random times would indicate Place Ville Marie is by far the most heavily travelled.

As we move down the aisle, almost to where we started in Place Ville Marie, we arrive at an opening leading to a cocktail bar, with three restaurants fanning out from it. Our tour is finished and we emerge opposite two Cathcart Street exits leading to McGill College Avenue. We entered Place Ville Marie from the Queen Elizabeth Hotel under Dorchester Boulevard and we exit at grade level.

Life around the clock

The social habits of Montrealers were ingrained long before this glittering new downtown existed. Montrealers are a sociable people with conversation, good eating and drinking ranking high on their list of pleasures. Being an evening people, they dine late and at a leisurely pace, between seven and ten o'clock if the well patronised bars and restaurants are any indication. And of course, Montrealers share with most Canadians an interest in another form of recreation: shopping or merely store browsing.

When one sees the full restaurants in Place Ville Marie and the Queen Elizabeth Hotel, the convivial lounges throughout the complex and the teeming shopping aisles of Place Ville Marie and Place Bonaventure, and then remembers the ugly gash that dominated the centre of the city for almost 50 years (see page 76), one might well wonder what the inhabitants did to amuse themselves prior to 1962.

There were other areas of the city that were popular then and are still popular. Stanley Street between Dorchester and Sherbrooke, for example, offers a great variety of entertainment from the wholesome to the dubious. And the Mountain-Maisonneuve area, with its coffee houses, informal and sometimes colourful eating places, has long been a favourite after-hours rendezvous for young and not-so-young office workers. St. Catherine Street and Decarie Boulevard have their share of fine restaurants and St. Catherine Street still booms as a shopping district.

Montreal Municipal Tourist Bureau

Bustling St. Catherine Street looking east from the Peel Street intersection. Dominion Square Building in the left foreground is a well-known landmark.

There are people who keep the facilities of Place Ville Marie and the Queen Elizabeth Hotel busy almost around the clock. They may be visitors attracted to the city by its new facilities and many patrons are found among the more than 22,000 people who work in the complex itself. More Montrealers than ever before are finding it worth their while to go downtown and to stay longer when they get there.

The stores in the promenade close at six o'clock on mid-week evenings and the shopping crowds thin out. But the aisles are not deserted. The promenade makes a pleasant stroll from Dorchester to St. Catherine, there are movies too and restaurants to be visited and windows to be shopped.

Perhaps the secret of the success of the entire complex from a social point of view is not that it offers anything Montrealers did not have before but that it offers the opportunity for them to more conveniently and comfortably enjoy familiar amenities. Visitors there are and visitors there will be, but the main patrons are Montrealers.

Statistics of food consumption are almost meaningless because there are few integrated complexes containing such a variety of eating places with which valid comparisons can be made. Average annual consumption of beef in Place Ville Marie and the Queen Elizabeth from 1958, the year the Queen Elizabeth opened, to 1968 was 1,182,600 pounds, the equivalent of a herd of 7,500 head. Lamb has been the next most popular meal at about 106,800 pounds. About 900,000 pounds of potatoes are served every year.

Thirty thousand dinners are served every month, while each day more than 10,000 persons sit down to a meal of some kind, ranging from a 50-cent coke-and-hamburger snack to a $60 a la carte gourmet repast. The latter may be accompanied by music provided by resident musicians employed in the complex.

The food service staff in 1968 totalled 1,125 persons embracing a variety of accents and dialects that would do justice to the United Nations. And think of the dishwashing involved in serving 52,000 highballs a month, 14,000 cocktails and 30,000 dinners. The liquor bill was $1.6 million.

THE SHIFT FROM ST. JAMES STREET

What was central Montreal like before 1962? The business community for years had been more or less neatly compartmentalized into two distinct segments. There was the big-business and trading centre on St. James Street and the commercial centre along St. Catherine Street to the north.

Most new commercial development, dating as far back as 1949, had been in the St. Catherine Street commercial core. The city planning department, in a publication called *Centre Ville*, noted that in the years 1949-62 office space in the central business district increased 77% from 11.4 million to 20.2 million square feet, most of the increase being concentrated in the commercial core. The floor area of public buildings, such as utilities and government administrative offices, increased 31% from 3.6 million to 4.7 million square feet and hotel space rose by 53% from 2.7 million to 4.1 million square feet. This growth took place almost entirely in the commercial core. Large portions of the rental space developed in this period came on the market in 1962.

Fragmentation of land ownership in the old St. James Street core and the attraction of Place Ville Marie were major causes of the massive uptown shift by large corporations seeking elbow room. But other forces—involving people themselves—were at work as well. The offices, banks and financial institutions in the St. James Street area employed large numbers of girls and women as clerks, stenographers and secretaries. St. James Street was a minimum of eight blocks away from the shopping amenities. A 15-minute walk uptown and a 15-minute walk back left little time for shopping or eating in a one-hour lunch break.

Montreal Municipal Tourist Bureau

St. James Street facing west toward Place Victoria. The tall building on the left side of the street is historic 360 St. James, home of the Royal Bank of Canada before the bank moved its headquarters into Place Ville Marie on Dorchester Boulevard.

St. James Street establishments had long had a history of high turnover among female staff members. "There's no place to shop down here", complained the girls, and promptly left their jobs for others uptown.

Finally, St. James Street itself began to move uptown. The Royal Bank forsook its historic old headquarters at 360 St. James to become prime tenant in Place Ville Marie. Hydro-Quebec and the Imperial Bank followed. The jobs offered by companies taking space in these large buildings further depleted the ranks of St. James Street female workers.

The three new Dorchester Boulevard buildings offering space for rent threw about three million square feet of prime space on the market all at once. Up to that time, the city had been absorbing only about 300,000 square feet of new office space a year. Men in the leasing and development business in Montreal still express amazement at the rapidity with which this space rented.

Jobs were opening up uptown and with all the between-hours and after-hours facilities of St. Catherine Street and Place Ville Marie beckoning, the plight of St. James Street employers became worse. The men were never particularly concerned. A business luncheon, a legal brief, a stroll or a book might occupy them during lunch hours.

It was not until Place Victoria opened in 1965 that the exodus of female employees slowed to a trickle. Now the girls had what they wanted, good restaurants, pleasant shops, hairdressing services, a movie house, all within minutes of their desks. For those working in the building, one simply rode the elevator to the shopping concourse and there they were—40 shops to be studied, inspected, compared and patronized. For those working at other St. James Street locations, the situation was certainly much improved.

Commenting on the attractions of the new downtown, one girl suggested that Place Ville Marie, as a working environment, appealed mostly to the "slightly shallow, superficial" type of person, while another claimed that employers, especially in retailing, were able to pay lower wages because so many young people were anxious to work there. Others recounted that they were told how lucky they were or "how marvellous it must be" working in Place Victoria or Place Ville Marie.

Place Ville Marie employers point out that despite the prestige involved in working there, they would be unable to obtain suitable staff if they paid substandard wages. But the desire of so many to work in Place Ville Marie does enable employers to be more selective in their hiring. They are in a position to reject many applicants.

It is almost axiomatic that in any city, "downtown" will mean different things to different people. In Montreal, for example, a man in the import-export business would think of St. James Street as down-town. To the suburban housewife who makes weekly or monthly shopping forays into the city, downtown is where the big department stores are. Still others identify it in terms of entertainment facilities; to them, downtown is where the theatres and night clubs are.

This has not really changed. The separation of commerce from finance is now less defined, with two major banks having moved uptown, followed by lawyers, public relations firms and other pro-fessional people. But St. James Street is still predominantly finance. St. Catherine Street has not deteriorated as a shopping district and there is no evidence it is about to deteriorate. Planners qualify this assertion by noting that there have been some casualties in areas where St. Catherine Street development was already weak, perhaps among establishments at the eastern and western extremities and where com-mercial development was thinly spread through institutional areas. Thus the north-south development has had the effect of upgrading the central core by hastening the demise of already weak enterprises. It has added a new dimension to this core by drawing it together to form a more compact and viable entity.

Place Ville Marie sits squarely in the middle of the prosperity of both St. Catherine Street and Dorchester Boulevard. These two arteries exchange people in a steady flow and Place Ville Marie, either across the plaza or through the promenade, is the shortest and most pleasant route between them. And the Queen Elizabeth Hotel and Central Station are as handy as Dorchester Boulevard.

Perhaps the essence of Place Ville Marie's architectural concept is found in the definition of architecture that says it is simply the con-trasting of voids and solids. The arrangement of the buildings on the seven-acre block complements the plaza, as well as offering a sense of spaciousness and unity with the city outside. Thus the voids are allowed to seep into the surrounding streets. The absence of exterior columns in the main tower enhances this effect. Of course, there are as many ways of employing the basic definitions of architecture as there are architects. At Place Bonaventure, for example, great areas of space have been completely enclosed in a sprawling, low-slung building covering six acres.

In spring and summer, thousands of Montrealers gather in Place Ville Marie's square every day to stroll in the sun, to people-watch, to attend parades, bazaars, political rallies, social and charitable events, band concerts and theatre performances. In the late evening, they may simply enjoy the atmosphere evoked by the glow of the clusters of Parisian-type lamps that dot the square. In front of one of the buildings along the west side of the plaza, is an outdoor cafe. Around the base of the cruciform tower are stores and shops.

In winter, the plaza does not fare so well. At the worst of times, the snow-laden north wind howls down McGill College Avenue and round and round the square, thundering from one building to another as though unable to find the way out. At the best of times, there is little to attract people.

The question is often asked why there is no skating rink or some other winter attraction. The necessary hardware to create a rink in winter and a great fountain and pool in summer were installed under the plaza when it was built. A skating rink was operated for a short time but was not successful. Some feel this was because an admission fee was charged, but Place Ville Marie officials say the intensity of the wind in the square discouraged people from using it.

But beneath the plaza, and reached through four large stairwells or court areas, is Place Ville Marie's shopping promenade. Here, weather ceases to exist and the promenade is busy all year round.

PLANS . . . AND MORE PLANS

While formal planning did not start until 1956, planning for Place Ville Marie can actually be said to go as far back as the mid-1920's when Sir Henry Thornton was president of Canadian National Railways, the owner of the land on which it was eventually built. Sir Henry himself had developed plans and even at this early time, they included the basic principles of a multi-level downtown core.

Montrealers well remember when the site was a big hole in the ground, an unsightly cleft referred to by Montrealers as "the Dorchester Pit". This ugly wound in the heart of the city was created about 1910 when Canadian National blasted a tunnel under Mount Royal and Dorchester Boulevard en route to what was known as the Tunnel Station, later to be replaced by the present Central Station. A flimsy-looking bridge carried Dorchester Boulevard across the cut, which extended for about two blocks in either direction. For half a century the eyesore remained. And as railway operations became more complex and more trackage was laid, the pit grew larger each year.

From as early as 1913 there were many development proposals. Sir Henry's in the mid-1920's were certainly the most promising but they collapsed in the market crash of 1929. But Sir Henry did accomplish something of considerable significance to future development. He initiated construction in 1929 of the existing Central Station which now

Upper: Canadian National.   Lower: Montreal Municipal Tourist Board.

Winter 1956: the "Dorchester Pit" viewed from the twentieth floor of the
Queen Elizabeth Hotel, then under construction on Dorchester Boulevard.
By 1962 the pit was completely covered by Place Ville Marie's seven-acre
plaza and its multi-level development, from which rises the cruciform tower.
In 1967, the plaza was enclosed on the west side by the IBM building. Note
the widened McGill College Avenue in the background.

plays such an important part in the subsurface street system.   The
depression forced a halt on this project in 1931, however, and work
was not resumed until 1939 with completion in 1943.

The importance of the Dorchester Pit site had long been recognized. It
was only a block away from Dominion Square, a short walk from
bustling St. Catherine Street and Windsor Station and after 1943, was
within stone-throwing distance of Central Station. Canadian National
had resolved that when development was finally undertaken, the entire
area would be developed as a single entity. Thus proposals that would
have filled it with a hodge podge of piecemeal, mediocre office construction were strenuously resisted.

When Donald Gordon, an executive with years of experience as Governor of the Bank of Canada and head of the Canadian Wartime Prices and Trade Board, became president of the railway in 1949, he made two significant decisions. The first was to build the 1,200-room Queen Elizabeth Hotel on Dorchester Boulevard across the street from the Place Ville Marie site. This was completed in 1958. The second was to do something about the Dorchester Pit.

It was Gordon, through an intermediary, who approached William Zeckendorf, a colourful and aggressive New Yorker with an international reputation as a developer and president of the American development firm of Webb and Knapp Ltd. In 1955, Zeckendorf was offered a 99-year lease on the site providing certain conditions were met. Canadian National had to approve the character and quality of the design, the developer and builder were to supply all the money necessary to build it, and a minimum of $250,000 was to be spent on research and design.

Zeckendorf called in the New York architectural firm of I. M. Pei & Associates and prepared a grand design for the entire 22-acre area bounded by St. Antoine, Mansfield, University and Cathcart Streets embracing Central Station and the Queen Elizabeth Hotel and the sites of the future Place Ville Marie and Place Bonaventure. The plan was approved by the railway in 1957 after Zeckendorf had spent more than $400,000 on design and research studies. Zeckendorf leased only part of the area, the seven-acre Place Ville Marie site.

Webb and Knapp (Canada) Ltd., Zeckendorf's Canadian subsidiary, began the project out of its own resources. An interim, or construction, loan of $50 million was arranged with a group of United States banks and at the end of 1962, Metropolitan Life Insurance Company issued a $50 million permanent mortgage to retire the interim debt. The remainder was equity capital provided by Webb and Knapp and two British companies.

Aside from financing, there were other difficulties. Tenants were needed to occupy the acres of office and retail space the complex would provide. Zeckendorf was admitted to be a wizard in real estate and financial circles but this kind of project was relatively untried. And to the layman, Zeckendorf's style appeared somewhat flamboyant and risky. Potential tenants were reluctant to commit themselves to a project that might never be completed.

The missing link was found when Zeckendorf met James Muir, then president of the Royal Bank of Canada. The Royal Bank became Place

Ville Marie's prime tenant, occupying seven floors in the main building and giving the building its name.

Following the bank's commitment, available rental space began to fill up. Among early office tenants were Aluminum Company of Canada, Air Canada, Montreal Trust, Canada Iron Foundries and Foundation Company of Canada, at the time Canada's largest construction firm with annual volume of about $150 million. (This firm has retained its name but was purchased by Janin Construction Ltd., Montreal, in 1968.)

Place Ville Marie has been called the most successful commercial complex in the world, and space rental rates would seem to back this up. Rock-bottom rates in mid-1969 were in the neighbourhood of $20 to $25 a square foot compared with $11 along St. Catherine Street, about $6 along St. James Street and $10 to $12 in the Toronto-Dominion Centre in Toronto. Leases are complicated things and no two are identical because rates charged vary with the type and size of business. Basically, they call for a percentage of gross revenue or a flat rate a square foot, whichever is higher. A revenue percentage equal to $40 a square foot is not uncommon in Place Ville Marie.

Why was Montreal the city in which this major multi-level city experiment occurred?

The prime prerequisite for such development is land, large contiguous chunks of it, preferably under single ownership. Montreal had such land conveniently located next to almost everything of importance in the city. The relationship of this land to the facilities of the city, such as the large department stores, the railway stations and the main business areas of both St. Catherine Street and Dorchester Boulevard, may be considered to be fortuitous. But on the other side of the coin, the CN had been holding the land for just such a development for half a century.

McGill College Avenue provides the perfect vehicle to carry the system northward to the busiest section of St. Catherine Street, and still further to the north but within reach of the system, are McGill University and Mount Royal.

Without the happy combination of circumstances that favoured Montreal, a series of year- or decade-consuming steps might have been necessary. First, the developer would have had to approach each small landholder in an effort to purchase his property. Secrecy would have been necessary to prevent prices from skyrocketing. Developers have been known to work 10 years or more through many different real estate brokers or through many different companies formed especially for this purpose, to assemble sites.

Holdouts, those who refuse to sell or will sell only at exorbitant prices, may be absentee landlords of the pockets of ramshackle housing found in many otherwise modern city centres. These landlords may receive high rents for this decrepit housing while spending little or nothing on repair and maintenance.

The developer himself may very well be one of the absentee landlords. If his site assembly has been partially completed, he may not want to spend money on maintenance when he intends to demolish the buildings shortly. Thousands of low-income families may be caught in the squeeze between the landlord and the developer.

An alternative to the procedure outlined would be to find land away from the city core but close enough to be grafted on to it. This is, in effect, what is planned for Toronto's Canadian National-Canadian Pacific Metro Centre development on 190 acres of waterfront land announced late in 1968.

The first $25-million element of this project, a communications tower, was expected to get under way in 1970. Total value, including a new transportation centre, hotel, convention centre, underground shopping malls, office buildings and apartment housing accommodation, is estimated at $1 billion. The project is to be completed over a decade or more. A separate $200-million project further east on the Toronto waterfront will consist of more apartment and office buildings and was also expected to be started in 1970.

In Montreal, rather than the developer having to wrangle with a multiplicity of landowners, the railway went looking for a developer.

The Royal Bank was in need of new quarters and so were other large companies in the old city. But St. James Street consisted of 30- to 50-year-old buildings of assorted shapes and sizes huddled shoulder to shoulder on both sides of the street. A firm requiring new quarters would have had to demolish the old and expansion would have been difficult or impossible. Splintering of land ownership—many owners each holding a small parcel—had developed, creating site assembly problems. This was in contrast to the situation on Dorchester Boulevard where, at this time, there happened to be land available in addition to that held by the railway.

Much of this additional land was occupied by somewhat unprepossessing structures, the owners of which may have been encouraged to sell by the dramatic nature of the development taking place around them. The land occupied by the old Gate House, a market produce building, happened to be up for sale when CIL went land shopping. The Bank of Commerce bought land from the Windsor Hotel at Dorchester

Boulevard and Windsor Street. About half the hotel structure was demolished to make way for the new skyscraper. Hydro-Quebec bought a lumber yard for its new building.

All this sudden growth in downtown Montreal created a serious transportation problem, with thousands of persons streaming back and forth to work each day. Discussion of the need for a subway system had been going on for as long as the Dorchester Pit had existed. Now the need became urgent.

The buildings that had created the problem provided the means of solving it. The new millions in tax revenues they generated were an important factor in enabling the administration of Mayor Jean Drapeau to borrow funds to start the subway. The subway, in its turn, was important in helping Montreal obtain Expo 67 and as Vincent Ponte notes, Expo in turn sparked massive highway improvements, a programme of expressway construction that normally would have been budgeted over two decades being completed in four years.

FROM PLAN TO REALITY

Variety in styles

One of the most distinctive features of the new downtown is the variety of textures. There is the slate-grey ruggedness of the Bank of Commerce Building, the corrugated smoothness of Chateau Champlain, the sleekness of CIL House and Place Victoria, the jumbo-type roughness of Place Bonaventure and the shimmering vastness of Place Ville Marie. None of these buildings is identical or even similar in texture, colour or shape.

In the centre of Dominion Square is a bust of Sir John A. Macdonald on a pedestal and covered by a concrete canopy. The unusual half-moon windows of Chateau Champlain across Lagauchetiere Street were derived directly from Sir John's canopy. Montrealers, either irreverently or affectionately, call the building the cheese grater. These windows have also been referred to as the "raised-eyebrow windows".

One professional critic refers to the "insular" character of the buildings which he attributes to their "inherent magnitude" and their "outward appearance as architectural monuments independent of their immediate surroundings." He adds that their links with surrounding buildings are not revealed in their physical form but are carefully hidden, at least from street level.

This comment may point up an issue of personal or artistic preference. One architect may prefer a measure of uniformity or at least a series of common elements while another prefers contrast and variety.

Canadian Pacific

The old and new stand side by side. Chateau Champlain is flanked by Mary Queen of the World Cathedral and CP Rail's Windsor Station.

Henry N. Cobb, partner in charge of the Place Ville Marie project for I. M. Pei & Associates of New York, architects on the project, notes the "compelling relationship" of the complex to Mount Royal to the north and says it would be impossible to overstate the significance to Place Ville Marie of this single attribute of the site. In other words, this was an attempt not only to change the shape of downtown but to give it a significance quite apart from aluminum and steel, to open a dialogue between the city and the mountain.

Cobb calls the site "a zone of maximum exposure, an intrinsically public place whose active participation in the mainstream of civic life was irrevocably preordained" and adds: "This all-important fact has occupied our consciousness from first to last and lies at the root of the design in all its aspects."

Cobb views the plaza itself as more than just a forecourt to a monumental building. It is a major protagonist in the confrontation of city and mountain and is the principle medium through which Place Ville Marie responds to its environment.

Fundamental to the concept was that the plaza should be a plane of strength that would limit, rather than be limited by, the buildings that rise from its surface. Consequently, the plaza was allowed to flow around the buildings, extending to the site's outermost boundaries, and under and through the main tower.

Place Ville Marie has been viewed and studied by architects, engineers and ordinary Montrealers from every possible angle. Cobb notes that the designers were concerned with distributing the building volume in such a way that the big tower itself would actively participate in the space rather than overwhelm it. The four slender projecting wings achieved this effect, says Cobb, and had the supplementary benefit of reducing the bulk of the building as viewed from any point in the immediate vicinity. But as many Montrealers have noted, this does not always hold true for more distant views when sun and shadow are absent. Under these circumstances, the wings lose their definition and the building looms bulkily on the skyline.

Place Ville Marie

Perhaps the main feature of Place Ville Marie as a construction project is the sheathing of aluminum curtain wall on the main building. The building is described by the men who worked on it as "a huge curtain wall job". Perhaps the simplest definition of curtain wall is that it is a light structure designed to keep the weather out and the warmth in. It has no load-bearing qualities of itself; that is, it does not support any weight and is fastened directly to the skeleton of the building.

More specifically, curtain wall is a two-layered panel of rigid material —it may be concrete, wood, steel or practically any other material—with holes filled by glass which become the windows. Between the layers of material is the insulation, in this case plastic foam. Place Ville Marie's curtain wall is of aluminum five inches thick including the insulation, and it holds panels of glass seven feet tall, five feet wide and one-quarter of an inch thick.

The skeleton of the Royal Bank tower is garbed in 14 acres of this sparkling glass and aluminum (about 90% glass) and a group of companies worked for 1½ years to develop this particular curtain wall system for Place Ville Marie.

Tests lasting up to a year in laboratories and wind tunnels preceded its installation. A major purpose of these tests was to find a way to keep rain, wind and dirt from penetrating the curtain wall through joints between the aluminum members and between the aluminum and the glass. This had been a recurring problem in aluminum curtain wall structures in the past. Wind tunnel tests subjected the wall to velocities of more than 100 miles an hour.

A special, rubber-like compound was finally developed to seal the joints as well as to separate the glass from the aluminum to prevent breakage of the windows. In the heat of summer, the aluminum surfaces may register temperatures as high as 120° Fahrenheit and in winter, 50° below.

This tower is a veritable showpiece of aluminum applications. There are 1,000 tons of it in the curtain wall, and another 1,000 tons in various other applications. Inside the building, electrical conduit alone uses 400,000 pounds. The curtain wall is mounted on millions of vertical members that if placed end to end would stretch for 22 miles.

The builders of Place Ville Marie were spared serious problems of excavation. Some trimming of the site was necessary but a minimum of digging was required to find rock capable of bearing the loads that were to be placed on it. There was already a hole in the ground and the track bed itself was solid rock of the required strength. Columns were anchored in the rock and construction was under way.

This was a very different situation than that faced by engineers and workers at Place Bonaventure, also built over railway tracks. Here, much difficult and hazardous excavation had to be carried out without the aid of heavy or sophisticated equipment.

In any building the size of Place Ville Marie, the complete structural framing is made up of two parts, the main grid system, which is what one sees when one looks at the unclad structural framework and the core or spine of the building in which the elevator shafts are located. The core of the building would normally require closely grouped columns and be reinforced by concrete sheer walls.

The Royal Bank tower's main grid system involved setting columns 25 feet 1 inch apart from centre to centre. The normal spacing would have been 25 feet but the extra inch was dictated by the position of the tracks. The tracks also prevented closely grouped columns in the core so the weight that normally would have been borne by these columns had to be transferred to the main grid system.

This weight transfer was accomplished by means of heavy trusses, horizontal members with diagonal bracings, that carried the weight to

the outer grid system. These trusses are 22 feet or two storeys high from train-top level to the bottom of the shopping promenade level.

The frame for the tower required 49,000 tons of structural steel supplied by Canadian, British and American mills. The Sun Life building, an aging but still attractive building next to Place Ville Marie, used 19,306 tons, the new 47-storey Bank of Commerce building 16,540 tons and CIL House 11,184 tons.

One aspect of this structural steel work illustrates the use of the computer in construction. The engineers wanted to calculate the amount of sway of the building under given conditions to enable them to select members of the proper strength for various uses. The members vary in thickness from one to three feet and some were among the heaviest ever fabricated to that time, weighing up to one ton a lineal foot and designed to support loads of up to 6,000 tons.

To illustrate what the engineers sought, imagine the Royal Bank tower being cut into horizontal slices, one for each storey of height. What would be the effect of certain wind loads on each slice individually and on all taken together? First, it was necessary to find the degree of movement of the top slice, then the second, the third and the fourth and so on to the bottom. But complications set in immediately below the top slice. The second had its own movement plus that of the one above it, the third moved in its own way with the additional movement of the previous two slices.

The thought of a 47-storey tower of glass, aluminum and steel rocking tipsily in the centre of a great city is enough to give one motion sickness. But, of course, it is not likely to happen. The engineers set the problem in a system of 128 simultaneous equations and turned it over to a computer. The formula was sound, the members were appropriately selected and the tower has only the necessary controlled sway of any tall building.

The four quadrants filling the right-angle spaces created by the wings of the tower are not connected to it structurally. Each of the quadrants has its own independent structural steel framework and is joined to the tower merely by glass and masonry. The 25-foot-high stonework facing topping the quadrants extends through the building almost to the core. From inside, one has the impression that four massive holes were left in the tower when it was built and the quadrants simply fitted into them.

Cantilevering permits a 16-foot column-free overhang around the building proper and 25-foot overhang of the quadrants. (In cantilevering, instead of providing support for horizontal members at either end, the vertical, load-bearing members are moved toward the centre.) The cantilevering also required extensive testing and computerized calculations. There will always be a degree of what is known as "deflection" (bending of the overhung members) but it is the extent of deflection that is important. Whereas an eighth of an inch would not so much as crack the masonry inside, a foot would bring down tons of rubble about one's ears.

In addition to eliminating exterior columns, the cantilevering also allowed a reduction in the number of columns required inside. Six columns were used for every 8,500 square feet of floor space. Column-less areas as large as 4,000 square feet are available on the lower floors and undisturbed areas of 260 square feet on the uppermost floors.

Place Ville Marie represented a new multiple tenancy concept in that it was an office building designed to accommodate major corporations, each of which might reasonably be expected to have a home of its own. The cruciform shape offers tenants natural light advantages not available in conventional rectangular buildings. Despite individual gross floor areas as large as 40,000 square feet, no desk need be placed more than 40 feet from a windowed wall.

Place Bonaventure

Place Bonaventure is built entirely of concrete. The material, used structurally, architecturally and ornamentally, is elephant grey on the outside and an aggregate-powdered rosy hue on the inside. One of the world's largest all-concrete buildings, Place Bonaventure is the second largest commercial building in the world with an area of 3.1 million square feet. (The largest commercial building is the Merchandise Mart in Chicago, designed for the same use as Place Bonaventure. The largest single office building in the world is the Pentagon in Washington, which covers 34 acres including a five-acre centre court.)

Winter construction is so common in Canada now that it is taken for granted. But this was not so a few years ago and many northern areas of the United States still lag far behind Canada in winter construction efficiency. The problem is especially serious where concrete is concerned because the water in the material freezes and will not set properly. The Place Bonaventure construction team draped tarpaulins and plastic sheeting over scaffolding and steam heated the resulting enclosure.

One of the contractual stipulations with the CN at both Place Ville Marie and Place Bonaventure was that train service was not to be interrupted while work was carried out on the substructures. But Place Ville Marie engineers were allowed to close down eight sets of tracks, funnelling traffic through the remaining eight. At Place Bonaventure, closure of only two sets was permitted.

In normal building construction, the contractor digs a big hole in the ground with high-powered, high-capacity equipment and then erects his foundations and columns. With the 16-track gorge running through the site and a useless load-bearing surface, a separate hole had to be dug for each column entirely by pick and hand shovel. Width between the tracks was 10 feet but the overhang of railway rolling stock took up most of this space. Thus to the height of the railway cars, columns could be no more than 3½ feet in diameter; above that height, they ballooned to stouter proportions.

Since there was no room for heavy equipment, men had to dig by hand 50 to 70 feet through earth, glass and stone to solid rock. Steel cylinders called caissons were sunk as the men burrowed into the ground. While the men worked, the caissons were lined by plank sheeting held in place by steel rings. The men entered the tubes, some no more than three-and-a-half feet wide, filled up their buckets, which were then winch-raised to the surface. The caissons were expanded at the bottom to the circumference permitted by the bearing capacity of the rock and were filled with concrete to carry the building loads into the rock. Despite the rigours of working underground, the cramped quarters and trains whizzing by only inches away, there were no accidents during this phase of construction.

Compounding the problems of substructure construction was the decision of the City of Montreal to run the subway line directly under the building. The decision was made after work was well started on Place Bonaventure, so for the contractor the difficulties were unexpected. The subway line came within inches of the foundations of Place Bonaventure.

The basic layout of the building called for concrete columns on a 25-by-25 foot grid. But the plan for a huge five-acre exhibition hall above the shopping concourse requiring large open areas necessitated a load transfer similar to that found in the lower levels of Place Ville Marie, except that this one was by design rather than being dictated by circumstances. The transfer was handled by gigantic concrete columns the equivalent of 1½ storeys of height. At the top of each, giant arms spread out like branches of a tree. These columns, set on a grid 50 feet by 75 feet and spectacular in appearance, serve both an architectural and a structural function.

Place Bonaventure is 17 storeys or 274 feet high. The components of the building are a five-floor merchandise mart of one million square feet providing showroom space for 2,100 manufacturers; more than 300,000 square feet of exhibition space including the 200,000-square foot Concordia Hall; an international trade centre of 55,000 square feet in which foreign countries exhibit products available on the export market; a 55,000-square foot Better Living Centre featuring displays of products for use in home and office; 350,000 square feet of retail space on two levels; and the roof-top hotel.

Sasaki, Dawson, Demay & Associates

Man and nature in a bustling urban centre: Place Bonaventure's roof-top garden.

In 1969, Great-West Life Assurance Company of Winnipeg, the firm which had provided most of the money to build it, took over sole ownership of Place Bonaventure and appointed Trizec Corporation, owners of Place Ville Marie, to manage it.

Place Bonaventure had become embroiled in a tax dispute with the city, and the complex was reported to be for sale for taxes. There was no question of Place Bonaventure's ability to pay the bill, reported in excess of $1 million. The management contested the bill and the protest was carried on by Great-West Life.

Early in 1970, there were parts of the project still incomplete. But Mr. James Soden, president of Trizec, was optimistic about its future. He says that when the market for its services had been created, Place Bonaventure will be a valuable asset to both Canada and the city of Montreal.

Place Victoria

When Place Victoria was built, it was the tallest reinforced concrete-framed building in the world. But such records are fleeting in construction and the probability is that somewhere in the world, there is a taller one. At 624 feet, it is the tallest of Montreal's skyscrapers but is still far from the height of Toronto's 56-storey, steel-framed Toronto-Dominion tower at 740 feet. Still taller at 784 feet will be Commerce Court being built by the Canadian Imperial Bank of Commerce, next to the Toronto-Dominion complex.

Place Victoria is another aluminum curtain wall building but the aluminum is painted black. To have had both Place Ville Marie and Place Victoria sun-sparkling in bright aluminum would have been tantamount to two *grande dames* meeting at a ball in exactly the same outfit. Place Victoria's black paint blends well with other colours and textures on the Montreal skyline.

While Place Victoria was under construction, lively controversy arose as to the suitability of concrete as a structural material for tall buildings, especially in view of Montreal's location in an earthquake zone. Place Victoria is the only concrete-framed building in Montreal that can be called a skyscraper in any real sense of the term and a rare exception was made by the city building department in issuing this permit.

Steel is the traditional structural material and a California engineer, J. Degenkolb, addressing a meeting of the Montreal branch of the Engineering Institute of Canada in 1962, declared it was the only suitable material. He said that in Los Angeles, another high-intensity earthquake area, bylaws had been passed to prohibit structural use of concrete in buildings more than 12 storeys in height.

A Place Victoria official replied through the press a few days later that Place Victoria's structural design had been tested by dynamic analysis, that the results of the test were incorporated in the design, and that the building was as earthquake-resistant as any steel-framed building.

In fact, a unique earthquake-resistant framing system was developed especially for Place Victoria by Montreal engineer John Barbacki, who worked with Italian architect-engineer Pier Luigi Nervi on the design of the project. The heart of the system is a huge X-shaped reinforced concrete wall running the full height of the building. This wall is linked by concrete trusses to the four massive bell-shaped concrete columns at each corner of the building. Stresses from earthquakes or high winds are met by the corner columns and absorbed by the core wall.

This framing system is visible to the casual glance and contributes to the sleek, streamlined appearance of the building. The smooth-rising concrete pillars relieve the blackness of the curtain wall and the trusses linking them to the core interrupt the exterior skin at the fifth, nineteenth and thirty-second floors.

In framing a building in concrete some steel must always be used. Concrete alone is a brittle material that can stand hardly any stress. Reinforcing steel—small-diameter rods of varying sizes depending on the strength required—run through the concrete to give it the required tensile strength.

Place Victoria probably sparked more interest among architects, engineers and students than any of the other major new buildings. Throughout construction there was a steady flow of such visitors to and from the site. Structural concrete was in fairly wide use in Europe but had been used only sparingly in North America. European buildings of the skyscraper type tend to be generally shorter than their North American counterparts. One of Europe's tallest is the 32-storey Pirelli skyscraper in Milan, a reinforced concrete structure also designed by Nervi.

But perhaps it was Nervi's reputation more than anything else that attracted attention to Place Victoria. World renowned as a "poet in concrete" or "virtuoso in concrete", in Rome Nervi is an architect, engineer and builder. In other words, he is a one-man construction team. Nervi is also famous for his philosophy on concrete and architecture. He refuses to be called an architect on the grounds that great architecture is basically a triumph of great engineering. So he describes Gothic architecture as "the product of incredibly deft and delicate engineering, amazing the eye with its applied ornament and structural logic."

He attributes to concrete a positive personality: "patient, strong, accommodating, willing to perform whatever tasks are required of it, proud to demonstrate its versatility; the finest construction material man has found to this day."

British architect Sir Basil Spence, rebuilder of Coventry Cathedral after World War Two and designer of Britain's Expo 67 pavilion, on a visit to Montreal after the building was completed called Place Victoria "the most beautiful skyscraper in the world".

While there has already been movement from St. James Street north to Dorchester Boulevard, there now are indications that Place Victoria will provide both the strength needed to keep the area from deteriorating farther and the link between the commercial core development and the westerly moving financial centre. Recent evidence of Place Victoria's stimulating presence is a $30-million, 32-storey skyscraper on Place d'Armes, three blocks east of Place Victoria, by Banque Canadien Nationale. This structure, sombrely handsome in its sheathing of black slate, has a modest shopping concourse beneath it with boutiques, rather than stores.

MONTREAL AND RESIDENTIAL CONSTRUCTION

This is the Montreal of the tourist and the downtowner. Residential construction in the city has not kept pace. Of the 400,000 housing units now standing in Montreal, 40,000 are ripe for immediate replacement. But, of course, as city housing officials readily admit, the word "immediate" is largely academic because there is no way this can be done. In addition, 70,000 need restoration or renovation to bring them up to modern standards. And after all this has been accomplished there should be another 40,000 units to accommodate families requiring housing and to maintain rent stability.

Montreal has its slum areas but many of the districts requiring demolition or restoration are not slums. They are dwellings built in the 1930's or earlier which do not measure up to modern standards. Among their defects, the doorways may be too narrow to admit such kitchen appliances as refrigerators and stoves, there may be little or no back-yard and there is no place to park a car. Fifty per cent of the homes in Montreal proper have no hot water and 20% have no shower or bath facilities.

In many cases, the people living in these houses are not poor and their housekeeping is beyond reproach. They do not live there because they cannot afford something better but because the supply of housing in the city is low and by choosing to save on rent they can afford other amenities; a car, a boat, a trailer and perhaps even a cottage in the country.

The problem is how to deal with substandard older houses in built-up areas of the city, with neither landlord or tenant prepared to take action. The onus is clearly on the city. Montreal has never had any deliberate anti-housing policy. But Pierre Bourgault, chief of the city housing department, says there never seemed to be the necessary money.

The money required to administer public housing is not always recoverable from tenants. For example, a family may be able to pay no more than $50 a month rent while it may cost the city $100 to administer that unit.

Incentive to action came in 1967 with formation of the Quebec Housing Corporation, a provincial equivalent to the federal Central Mortgage and Housing Corporation. The city housing department formed in November 1968 enabled the city to participate in federal-provincial shared-cost housing programmes.

Under this arrangement, the federal government pays 90% of construction costs and the province and city 5% each. The province picks up 75% of the administration deficit, and the city 25%.

But city housing officials say they are having to run hard just to stay in the same place.

We have seen that assembling a site for Place Ville Marie and Place Bonaventure were not serious problems because the land had been sitting vacant for half a century anyway. There was no need to demolish houses and move people in order to bring about the multi-level city. Still, this glittering new downtown may be said to be advancing on residential Montreal. As in Toronto, Vancouver and many other major Canadian cities, hundreds of high rise apartment buildings are being erected for which it is necessary to demolish units of reasonable rent.

In 1965, Radio Canada, the French language arm of the Canadian Broadcasting Corporation, announced plans for a new headquarters building in south-central Montreal. The contract for the foundation work was awarded in 1966 but the project still became something of a political football. Work proceeded on a stop-and-go basis although early in 1970 the superstructure had reached the fourteenth floor. About 1,000 housing units were cleared from the site when the building announcement was made. More than 12,000 inhabitants were displaced from this site and other areas required for construction of Decarie Boulevard, a major east-west expressway completed just before the opening of Expo 67.

The housing department has since set up a social work department to aid in the resettlement of families displaced by construction and urban renewal programmes.

In 1969 the housing department and city planners felt that no housing in the city should be demolished, regardless of its condition. Units planned for demolition would be upgraded and made to serve as stop-gap housing until better could be supplied. Aside from the 770-unit Jeanne Mance project, built in the mid-1950's just a stone's throw from Place des Arts in east-central Montreal, no new replacement housing has ever been built in the city. Most of the housing stands as originally built and in Jeanne Mance, exactly as many units were constructed as were demolished to make way for it. The net gain: zero.

This does not mean, of course, that no houses were built. Through the years, there were new houses being erected on new sites by private individuals and developers.

The two new city housing projects are Operation 300 and La Petite Bourgogne. Operation 300 will spot 364 units in four corners of the city proper while La Petite Bourgogne, a 10-year programme, will see eventual construction of 4,000 units to accommodate 16,000 persons.

La Petite Bourgogne is in a residential-industrial area bounded by Atwater and Guy Streets, the Canadian National Railway tracks and the Lachine Canal. It is little more than a mile from Place Ville Marie. The area will be all residential and this will require removal of seldom-used railway tracks and extensive transplanting of light industry. No houses will be demolished that can possibly be restored, although there will be some casualties. Everything possible will be done to preserve the old world character of existing dwellings while making them functionally competitive with modern units.

Operation 300 will be all new units, town house style, and will provide room for car parking beneath them. The housing department plans to mingle fully self-supporting families with those who need rents geared to income.

The downtown Montreal concept is far more important and socially significant than the concrete and steel or the batches of aluminum that went into the structure. The St. Lawrence Seaway, for example, was a larger construction project. But Place Ville Marie and similar projects that are to follow in other parts of the world, affect people where they live, work and play, in the cities.

In the future, it is quite possible that some other city will be the vehicle for further development of the multi-level city. A major water-front plan proposed for Toronto could eclipse the Montreal effort. The city of Calgary is studying a plan which could put pedestrians on enclosed elevated walkways throughout most of the downtown area. And as we have seen, a number of American cities are progressing with plans of their own.

But, Ponte says: "However impressive such developments may be in scale, they are few and have not always been integrated with the surrounding city structure. They stand simply as islands of order amidst the surrounding chaos."

CHAPTER 3

Industry on the Move

Building for People and *Developing Water Resources* have examined a
number of major Canadian construction projects, how and why they
were undertaken and their impact on the lives of people. But the story
would not be complete without reference to the industry that produced
them. This chapter will outline the structure and organization of the
construction industry, its problems, the peculiarities that make it different from other industries, its growth and development and what the
future may hold.

STRUCTURE OF THE INDUSTRY

Picture an animated cartoon on television. In the centre of the
screen is a box-like building, a factory. Flowing from many directions
into one end of the factory are trainloads and truckloads of materials:
steel, rubber, glass, plastics, fabrics.

No people are visible but there is evidence of human activity. From
the opposite end of the factory emerges a steady flow of shiny new
automobiles. Watch the plant for a day, a week, or a year; it never
moves. The materials flow in and the product flows out.

Change the channel. Picture a huge vacant field. Then picture
trucks, power shovels, graders and a small army of people descending
on it. Before our eyes, a great gaping hole appears in the ground. Then
a network of steel or concrete girders and columns rises out of the hole,
eventually to be clothed in steel, concrete, glass, or aluminum walls. The
muddy soil around the building suddenly sprouts shrubs and flowers.
The men clean up their debris, pack up their equipment and exit.

A grasp of the idea illustrated here is essential to any understanding
of, or feeling for, the construction industry. The idea, in a word, is
mobility.

In manufacturing, and in most service industries, the plant and the
workers remain stationary. The manufacturer brings the materials to
his plant for the workers to fashion into useful or desirable products,

which then leave the plant. In construction, the people and the plant move. The product remains as a permanent feature of the landscape.

To this central fact of mobility may be added what might be called fragmentation and lack of product standardization. Individual construction firms rarely have control over product standardization or the kind of product they will be required to produce. The customer specifies exactly what he wants and it is up to the construction firm to produce it. And in each case, it is different. Can you imagine two more unlike projects than Ontario's Macdonald-Cartier Freeway and Montreal's Place Ville Marie?

(The only exceptions to the previous paragraph occur if a contractor has sufficient funds to finance his own projects. He may then play the role of the owner-builder—as is often the case with apartment construction—or he may have his own design team on staff who contract with owners in "package deal" arrangements. Such arrangements generally permit the construction firm to develop patented systems which lend themselves to prefabrication.)

There are giant construction firms and there are midgets. But most construction projects are carried out by one of thousands of small firms or units scattered across the country. They are little known outside their own immediate sphere of activity and their own circle of clients and suppliers. This is what we mean by fragmentation.

These three ideas—mobility, heterogeneity of product and fragmentation—make their presence felt in every corner of the industry, touching on technological development, the structure of the industry as a whole, the organization of individual firms and the manner in which a construction firm sets itself up to tackle a new project. There are no assembly lines and few automated machines. No substitute has been found for the man with the wrench, the bulldozer, the saw. Thus construction is the most people-oriented and most personal production industry of all. Roadbuilding could be an exception because of the large and specialized equipment used.

The people

Aside from office staff and a select group of highly skilled tradesmen, few construction firms have many permanent employees below the rank of foreman. Again, this is because of our trio of basic ideas underlying the structure of the industry. There are many small units, often with little capital, and they are on the move. The demand for their product is not always steady. These contracting firms can afford to employ only persons who are actually producing.

In a large company, the chief executive, along with the board of directors (if any) and other executives he may take into his confidence, are responsible for determining the firm's broad policies, establishing its objectives and directing its operation.

The chief executive also keeps abreast of labour legislation and labour relations; watches construction business forecasts as well as those for the general economy; studies trends in construction cost; and keeps up-to-date on tax changes and general legislation. He also needs to know as much as possible about developments in equipment and materials. Periodic visits to his construction sites, and reports from superintendents and managers, keep him informed of the financial and physical progress of each job and the overall financial state of the company.

It is to his staff that the contractor must look for the drive and skill to bring a project in on time and at a profit. The estimator, known as the contractor's inside man, starts his work before the contractor knows he has the job. In the field the project manager, or outside man, assumes command, often of more than one job at a time. Each job is supervised by a project superintendent. These men come in assorted sizes and temperaments and may have colourful personalities. Yet superintendents today often look and act more like vice-presidents than legendary muddy-booted front-line supervisors.

The responsibilities of the superintendent and the project manager are to watch costs; maintain the best possible relations with the owner's representatives (the architects and engineers on the site); submit reports to head office on costs and work progress; and most of all, keep the work moving. They note carefully segments of work that are ahead of or behind schedule. When a section is going well, it will not require so much of their attention, and they can concentrate on parts that are moving slowly.

Large construction firms may have professional managers (men who have trained in university for business management positions), or managers who simply moved up through the ranks but who do not own the company. But the typical construction firm operating today was founded by the man now heading it, or by his father. Construction has been called the last frontier of the old-style entrepreneur, although some inroads are being made by the professional manager.

This background has created certain contractor characteristics that are immediately identifiable: independence, individualism, a tendency to go it alone, ignoring security in the bid for a big payoff. Contractors are tremendously inventive in the field but often careless in administration. They may tend to concentrate so heavily on moving dirt or putting steel and concrete together at one end of their operation that they fail to see their profits draining away at the other. This, fortunately, is changing.

The construction firm gathers its labour wherever the project is located and in a few weeks can expand to a work force of thousands. On small urban projects, workers are obtained from the local unions. On major projects in isolated areas, a work force must be assembled and transported to the site, and may remain basically unchanged for five years or more. But the end is the same on almost all construction projects. The men collect their final pay cheque, have a party, and say goodbye as they leave for new jobs.

Construction defined

A distinction needs to be made between construction as an activity, and the construction industry. The total value of construction activity is what economists and statisticians look for when tallying up a year's construction volume or forecasting anticipated volume for the next year. But, in fact, a large part of this value is accounted for by activity on the part of firms or organizations whose primary business is not construction but something else, such as railways.

The major divisions of construction activity are industrial maintenance, building construction, and engineered construction. The second may be further categorized as residential or non-residential. Repair and new construction are other possible subdivisions: the new construction of today creates the market for repair jobs of tomorrow.

Residential construction covers any type of dwelling accommodation. Non-residential building may include industrial plants, offices, hospitals, shopping centres, schools and other institutional and commercial structures.

Engineered construction (also known as heavy construction) takes in roads, bridges, highways, airports, water works, sewage, systems, dams and irrigation works, electrical power projects, telephone, gas and oil facilities, wharves and marine works.

The line between building and engineered construction becomes almost indistinguishable when one considers the engineering intricacies of modern high-rise buildings and the downtown complexes appearing in our cities. In these projects, which might better be called architectural construction, the emphasis is on both structural design and aesthetics.

In both architectural and engineering construction, the engineer is a manipulator of natural forces; the architect is a manipulator of spaces and textures and, in the design of massive buildings, relies on the engineer to tell him whether his idea is workable.

Construction projects can also be classified according to the nature of their ownership—that is, whether they are publicly or privately owned. Almost all *engineering* construction performed in Canada (about 40% of the annual *total* construction value) is performed by or for governments, municipal, provincial or federal. Governments also account for a considerable proportion of *building* construction.

Broadly defined, the boundaries of the industry take in people in many businesses and professions who do not build: civil, mechanical and electrical engineers; architects; city and town planners; building supply firms and equipment dealers. If we extended these boundaries far enough, we would also have to take in surety and bonding companies; bankers who are frequently called upon to provide money; equipment-testing agencies such as the Canadian Standards Association; and research agencies such as the Division of Building Research of the National Research Council.

Narrowly defined, the industry includes only the people and organizations engaged in the business of contracting to carry out construction work on a full-time basis. They range from the relatively few national and international giants who take on the multi-million dollar jobs to the many one-man operations which renovate or remodel homes.

Tapping the market

The dominant figure in the construction industry proper is the general contractor. The term "general" has nothing to do with the kind of work that he does but refers to a method of contracting. He may do light industrial and commercial work; he may build dams or bridges; he may remodel homes.

When the general contractor assumes that title, he also assumes total responsibility for completion of any project he undertakes to the complete satisfaction of his customer. He responds to public bid calls inserted in newspapers or periodicals by various public agencies, or to invitations to bid extended by private owner-clients.

The "negotiated" contract and the invited bid are two procedures almost entirely restricted to work for private clients. The public bid call is one of the oldest institutions of democratic government. These bids are opened in the presence of the bidders or any other persons who care to attend. If the contractor is qualified and his price is the lowest received, he enters into an agreement with the owner.

Estimating costs is one of the most vital functions within the contractor's organization. An error on the part of the estimator can cost the firm thousands, even millions, of dollars. The estimator or the estimating department prepares the bid on which the contractor bases his hopes of obtaining jobs to keep his equipment and men working and of earning a profit.

Of course, the estimator may have functioned intelligently and cleverly and may still find himself just a hair's breadth above the low bidder. In such cases, his firm usually does not get the job.

From the time the contract is awarded, either on a bid call or negotiated basis, the general contractor is in charge. If there are mechanical and electrical sub-contractors to be hired, he calls for bids or negotiates in the same manner as the owner-client for the overall project. The general contractor receives all the money when due and pays the sub-contractors in accordance with their agreement. Job overhead is generally his responsibility and he allows for it in his bid. He co-ordinates the work of his own forces and that of the sub-contractors.

While general contractors may do any kind of work, they tend to specialize in one field, such as building or engineering construction. A few may do many different kinds of work. Road builders, engaged in construction of airports, roads and streets, are usually considered apart from general contractors because of the specialized nature of their work and the expensive specialized equipment required. Sub-contracting in road building is minimal.

Some jobs are too big for a construction firm to handle alone. A firm will then temporarily team up with one or more partners in what is known as a joint venture or consortium. Every joint venture has a sponsor, the firm which organizes the group and takes the lead in its affairs. Any firm finding a suitable job and knowing other firms equipped to take part can become a sponsor, yet the firm who takes the lead on one job may become part of a joint venture headed by another firm at a later time.

Design

The architect and engineer begin work on a large project years before it goes to the field. They are the professionals first approached by the owner when he decides he needs a structure. As agents of the owner, they may remain on site throughout construction.

On a building project, for example, the architect prepares a design, consulting with civil, mechanical and electrical engineers as necessary— often a lengthy process. On site, he inspects and approves the work, ensuring that it is performed in accordance with the contract and the specifications.

Despite what many people may think, the architect is not a dreamer closeted away in an ivory tower, conjuring up strange designs to bewilder or delight the eye. He must be a tough, realistic, practical businessman. On the construction site, he has the power to make binding decisions when disputes arise, and to determine compensation for changes in the work ordered by the owner while the job is in progress. He can halt the work completely if he feels it is not being done properly.

The architect also passes on the money due to the contractor and has the authority to withhold progress payments if he feels this is necessary to protect the owner from loss due to failure of the contractor to meet contractual obligations.

Many construction men are graduate engineers and find use for engineering talents in their daily work. Engineering is an invaluable background for estimating, supervising projects and other jobs in the industry. But when an owner or an architect wants a structural design for a mechanical or electrical installation, he goes to a specialist who is usually in private practice. In small buildings, there is no need for a specialist engineer.

Service provided by the construction industry generally has been satisfactory in the past, but with modern projects becoming larger and more intricate, owners are demanding even better service. They are allowing less time for construction and insist on earlier occupancy than is possible under traditional methods. The industry, or some parts of it, have discerned that the greatest time-saving possibilities lie in that lengthy period between commencing design work and the start of operations in the field.

To eliminate this time lag, a new form of construction management— its official name is just that, Construction Management—has been devised. Stripped of its complexities, it simply means that design is carried out simultaneously with construction. And, of course, it requires close consultation between designer and builder from the very inception of the project.

With jobs becoming more complex and owners wanting quick completion, better scheduling is necessary. Working from day to day "by the seat of one's pants" is no longer possible, particularly when competitors are studiously researching each element going into a project, assigning a time and cost figure to it. Detailed control techniques, some employing the computer, have been devised. The Critical Path Method (CPM) is probably the best known and most widely used of these advanced tools of the progressive constructor.

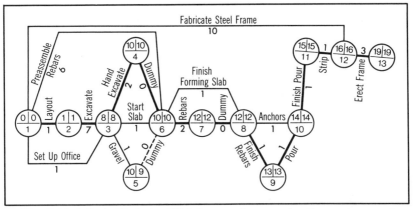

The Critical Path Method—CPM is a modern construction management technique which permits planning and scheduling using the logic of simple arithmetic calculated by hand or, for jobs of immense complexity (such as Expo 67), by computer. Interrelationships and duration of activities can be seen at a glance. The scheduler can pay special attention to those critical aspects which, if delayed, would affect the final completion time. Some jobs (such as Job 1-6, "Preassemble Rebars"), have several days of "float time" (in this case, four days). The CPM chart shows that that activity can be performed anytime between Day 0 and Day 10 without becoming critical and delaying the whole project. Critical jobs (shown on heavier line) have no float time.

The owner

The owner is the man (or organization) who pays the bills—the construction industry's link with the economy of the country. Through the owner vast sums of money are funnelled into construction each year.

Owners are found just about everywhere. Who are they?—the T. Eaton Company and Simpson-Sears, Algoma Steel and the Steel Company of Canada, General Motors and Ford, Shell Oil, and Imperial Oil, the chartered banks, and so on. They are all buyers of construction goods and services, all customers of the industry, not just on a one-time basis but on year-round, year-after-year construction programmes, either repair or new. Such firms may meet a part of their own construction requirements but at one time or another, all will buy from the industry.

School systems, churches, hospitals and charitable institutions are also buyers of construction. But the largest single group are governments: municipal, provincial and federal.

Many factors enter into an owner's decision to build a dam, bridge, power project or skyscraper. In the case of public works projects, the need may be self-evident or the project may be undertaken in response to public demand. In private construction, the prime consideration is whether the finished project will yield a satisfactory return on the money it will cost.

Construction and the economy

The construction business is one of Canada's largest. Construction consistently accounts for close to 25% of the annual gross national product: that is, the total value of all goods and services produced in the economy, from shoe shines to grand pianos to power projects.

In 1968, the construction industry proper paid out 2.6 billion in wages and salaries, compared with $9.1 billion paid out by all manufacturing industries combined. The January, 1970 issue of the *Labour Gazette,* an official publication of the Federal Labour Department, reported a total of 423,900 persons actively engaged in construction during July, 1969. During the peak periods of the year, usually about 600,000 persons are employed directly on site. Another 650,000 are engaged in related fields of manufacturing and distribution of building supplies and equipment.

THE PAST . . .

Well, John Henry said to the captain:
Oh, a man ain't nothin' but a man,
Well, let your steam drill beat me down
And I'll die with my hammer in my hands,
Lord God,
I'll die with my hammer in my hands.

A traditional folk song, the John Henry legend is based on fact. A 34-year old Negro, John Henry, swung a 10-pound steel-driving hammer inside the Big Ben Tunnel on the C & O Railroad in West Virginia in the early 1870's. The contractors had decided to try the new steam-powered piston drills, when Henry, the best driver on the Big Ben, fearful that his job was at stake, decided to challenge the new machine. There is some doubt as to whether or not Henry actually died from the ordeal.

While Samuel de Champlain in the early 1600's was the first large-scale builder in Canada, most of the construction we see around us has been put in place since the First World War. Champlain engaged in both building construction and what today would be called heavy or engineering construction. In about 1605 he commenced the palisades and habitations at the Port Royal settlement (now Annapolis Royal) in Nova Scotia. And if the Vikings and Indians erected buildings before he did, Champlain certainly became the first Canadian road builder about 1620, when he laid a 20-mile graded road from his settlement to Digby Cape.

There was little organized construction for many years after that. Settlers simply traded skills in building homes for themselves as they moved up the Ottawa and St. Lawrence Rivers. The materials were drawn from the plentiful supply of timber and stone. The first iron forge was not established in New France until 1733.

Actually, in the early part of Canada's exploration and settlement period, most construction was confined to townsites. There was little need for road-building or massive excavation, and little desire to pursue it in such rugged terrain.

Early construction was characterized by back-breaking, flexed-muscle manpower coupled with beasts: horses, oxen and mules. In addition, the crude methods available placed limits on the size of the excavations. For example, when the shoveller reached the centre of the excavation, how was he to throw the dirt out? And how was he to throw it out when the excavation got too deep? Some of these problems were never solved economically until mechanization came along.

With the exception of shovellers and pickmen, most of the construction workers from 1600 to the mid-1800's were skilled tradesmen, such as stonecutters, master carpenters and joiners. These men worked with raw materials at the site. Later, materials were finished elsewhere and delivered to the site only when they were needed for assembly into the structure. This traffic control has been developed to a high degree of perfection today.

The first major heavy construction works in Canada—forts, docks, canals—were all designed by military engineers (as distinguished from "civil" engineers) and built by soldiers. Such was the Rideau Canal connecting Lake Ontario to the Ottawa River at Bytown (now Ottawa). It was intended to provide a route to the St. Lawrence in the event the Americans should invade the Canadian provinces.

The early 1800's, however, saw great strides. Portland cement, the key ingredient in concrete, was developed in 1825. The reduced cost of producing steel made larger, more complex projects possible. The steam shovel was brought to a useful state of development in 1839, too late for use on the Rideau Canal which had been completed in 1826, but in time to help in railway construction from the 1850's. Canada's first iron bridge, Montreal's Victoria Bridge across the St. Lawrence River, was completed in 1859.

Mechanization in any form was the exception rather than the rule in Canadian construction until as late as 1900. However, one widely-used piece of basic equipment was the derrick hoist. This was simply an upright tree trunk with an angle boom projecting out from the base. Ropes and guy lines supported the boom, while a horse or manually-operated windlass did the lifting. The boom was usually pivoted for some rotation around the upright.

In lieu of mechanical equipment, blasting was used even more extensively in early construction than it is now. Material break-up that would be accomplished today by power shovels and tractor-drawn rippers had to be carried out by a seemingly never-ending series of small blasts to reduce the material to manageable muck. And until Alfred Nobel invented dynamite in 1866, blasting had to be carried out with dangerous, unstable, black powder.

The coming of steam

Railway building was (and still is) a specialized area in the construction field. It draws heavily on many branches: excavating for grades and right-of-way; masonry and concrete construction for bridges and structures; tunnelling and heavy rock work. The widespread use of the steam shovel in the railway construction proved the ability of a new source of power which was reliable and reasonably compact for the times. Steam changed construction industry procedures in a matter of years. The folk-song "John Henry", which tells of the steam drill overwhelming the spike-driving hero in the 1870's, reveals this sudden turn of events and preserves it in song. In the construction of Canada's first slender sea-to-sea transportation route (completed by Canadian Pacific in 1885), the steam shovel had a major role.

One CPR engineering triumph—the spiral tunnels deep within the Rocky Mountains—could not have been possible without the power of steam. But for smaller jobs, contractors still used methods that had not changed much in a century. Horses drew a scraper bowl across a site until a semblance of an excavation had been finished, then shovelmen dug out the edges.

Horse-drawn rigs served as the basis of design for several high-powered machines found on modern-day sites. Many of these implements were invented for farm use and quickly adapted by construction men.

CP Rail

Early railway building had to use large labour gangs. There was little mechanization.

The belly dump truck, for one, is the descendant of the small half-cubic yard bottom-dump wagons which worked on every highway and byway for over half of the nineteenth century. The ancestors of the modern scraper, grader and roller—even the bulldozer—were originally powered by teams of horses or oxen. Mechanical design has been improved—usually passing from horse, to steam, to gas, oil or electric power—but basic principles remain unchanged.

There are still occasions when the horse may be used to advantage. Construction of the nuclear research town of Pinawa, Manitoba, in the early 1960's was hastened when the contractor found that teams of horses could manoeuvre supplies around the muddy site more economically than his most modern tractor. Bush-clearing contractors, too, occasionally use similar economies. And even though use of the animals is now a rarity, it is not uncommon to find clauses in general conditions of standard contract forms or municipal by-laws which detail the required treatment and feeding of horses and teamsters.

Internal combustion engines

While steam power did become a key factor in large-scale construction operations, its massive bulk precluded its displacing the horse in the more common, smaller-scale projects. That honour was reserved for the internal combustion engines.

Development of smaller and more compact construction equipment was a by-product of the emergence of these gas and oil-fired engines and the related progress of the automobile just prior to the turn of the twentieth century. These lighter and smaller power plants, whose development was initiated chiefly in the United States or Britain, permitted construction machinery designers to offer equipment priced within the range of all contractors, not just the largest firms on the largest jobs.

Tractors, for instance, became important to the construction industry after making their mark on the farm. Henry Ford's Fordson tractor, available immediately after the First World War, was undoubtedly the most popular and certainly the cheapest. Almost as many Fordsons rolled off the assembly lines as Model T Fords. Such tractors served as the power units for the industry's earliest self-propelled graders, tractor shovels, front-end loaders and scrapers.

The crawler-tractor, too, made a big impact on the market after the 1914-1918 War. Invented in 1908 by Benjamin Holt (one of the founders of the present-day Caterpillar Tractor Company) to counter the soft delta mud of California orchards, the lay-your-own-road principle proved its worth and was developed considerably in wartime tracked armour vehicles, such as tanks and gun carriages.

On construction jobs, crawler machines produced new levels of job productivity and did, in fact, permit many jobs not previously feasible, especially when a variety of attachments such as bulldozers were added.

Based on an earlier horse-drawn version, the self-powered scraper was developed at this time by an inventive American, R. G. Le Tourneau. But his machine required two men—one to drive and steer and the other to operate the dirt-handling mechanism for the low-slung scraper bowl trailing behind. Le Tourneau later solved this problem by installing cable controls and, still later, hydraulic controls that permitted a single man to operate the machine. The scraper itself he converted to rubber-tired wheels with a goose-neck connecting the bowl to a two-wheeled prime mover. Under the competitive pressures of a dozen manufacturers and the construction industry's demands for more productive equipment, the scraper bowl grew from half a cubic yard to today's average of 35 cubic yards (some are as large as 80 cubic yards).

During the early 1960's, after more than 40 years' creative experience in the field, Le Tourneau devised still another invention which may yet revolutionize earth-moving to the same extent as his cable-controlled scraper. This invention was the electric wheel. Here an electric motor

An early two horsepower bulldozer.

The steam engine started a revolution in equipment design.

Construction equipment design, such as this scraper, owes much to farm machinery.

Dump wagons hauled rocks for early roads.

Westinghouse Air Brake Company

Building trans-continental railways in Canada would have been impossible without the steam shovel.

Remote-controlled fleets are not far off.

situated in the hub of each wheel provides them individually with power; no transmission is needed and rheostats control wheel speed electrically.

Parallel improvements to steam-powered shovels and draglines came during the early twentieth century. Diesel and gasoline engines were installed in these machines too, making smaller capacities feasible and marketable. (Smaller jobs had not been able to make economical use of the giant steam-driven rigs but could readily apply the smaller internal combustion ones.) This increased the machine's popularity and inspired development of modern-day back-hoes and truck cranes. But their steam-shovel ancestry is usually still apparent since basic changes have been minimal.

. . . AND THE FUTURE?

Major breakthroughs have not typified the progress made in construction techniques, equipment or materials. Therefore any prediction as to the future role and shape of construction can fairly safely be based on past experience. Yet one other characteristic facet of the improvement process within construction—the fact that many ideas and techniques originate in other disciplines, such as pure mathematics, chemistry, physics, metallurgy, agriculture, biology and the social sciences—means that revolutionary changes within any of these disciplines might very well upset construction's pattern of evolutionary change.

Role of the computer

Still, much of construction's future shape is predictable. The computer, for example, has already made its mark and will continue to make significant inroads, particularly in design. The day may already have arrived when a Canadian design engineer without access to computer aids must be considered as antiquated as a draughtsman using a quill pen.

Computers, unlike the human brain, cannot think independently. They must be programmed. But their plus factors are accuracy and speed. Many tasks hitherto considered too time-consuming or error-prone to be of practical value to the designer now can be performed in a matter of minutes. The result is imaginative and economical designs never before feasible. Buckminster Fuller's 250-foot diameter geodesic dome at Montreal's Expo 67 is a case in point. Here, 23,274 individual elements were computer-designed and dimensioned to distribute the loads on the structure, all without the reams of working drawings and calculation sheets which otherwise would have been required. The result was the most economical space enclosure ever built by man—four ounces of material to enclose each cubic foot of space.

One computer programme from France, already being marketed, can design highways at the press of a button. Permanent memory data includes all pertinent standards of road design, such as permissible grades, curvature and pavement widths. To design a road of that standard through a specific area, a contour map of the study region with such local restrictions as traffic volumes and soil types added to the input, is scanned electronically by the computer. Future sophistications will permit input data to be movie films shot obliquely over the terrain from low-flying helicopters.

United States Information Service

Architect Buckminster Fuller's geodesic dome is the most economical enclosure ever designed. Only four ounces of material enclose each cubic foot of space. The largest yet built was the United States Pavilion at Montreal's Expo 67.

The computer chooses the best possible route through the region, printing out all necessary calculations including survey notes for the field crew, alignments, grades, quantities of cut, fill, borrow and waste material, and the costs of each phase of the road-building process. It then produces drawings of the road's profile, plan and typical cross-sections. Finally, it assumes the position of a driver's eye proceeding along the new route, and draws perspective sketches of the terrain and road features seen at regular intervals. If this electronic eye does not like what it sees—for instance, a blind curve or a confusing intersection —it changes the design to eliminate the hazard.

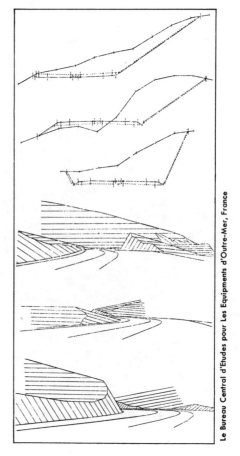

Computer-drawn sketches indicate what a driver would see while travelling along a computer-designed roadway.

A natural extension of this road-design programme is to integrate bridge-design calculations into the computer programme so that it will choose the best bridge locations, and the most economical bridge type, and then design it. Eventually—once the construction industry can handle the situation—this output will also include an outline of the equipment needed to complete the job; all the necessary order forms for concrete, steel and asphalt; even detailed shipping labels for the reinforcing steel rods.

To utilize such computer capabilities, the construction industry must make a basic transformation; it must permit itself to become a more internationally standardized and more monolithic industry than at any time in its history. That is a difficult proposition for an industry typically made up of individualists.

But examples of such change are already in evidence, particularly in the building industry, where standardization and co-ordination of dimensions are rapidly catching on, resulting in more efficient projects. One such example is modular (or dimensional) co-ordination, which required invention of a new dimension, the module, abbreviated M. Internationally, one M was set at 100 millimeters; but North America held out for an M length of its own, four inches, which is close but not quite equal to, the metric standard. But even this efficiency move often has to be imposed on building material manufacturers by purchasers—government or other project owners who want a more economical job.

New materials

Yet the design element is not the only evolving aspect of construction. Materials, too, are constantly being improved, or tried in combination with each other, to produce a better, more economical result.

Steel and its alloys provide excellent examples. Metallurgical science, a field construction draws heavily on for improved materials, has provided numerous combinations of the basic iron/carbon mixture which makes up steel. To them are added minute percentages of other elements such as nickel, manganese or titanium, thus transforming the metal.

As a result, new metals meet new needs: a Canadian-developed grade of steel able to withstand sub-zero cold without becoming brittle to the point of failure; ultra-high strength steels, never before thought possible, for use in pre-stressed concrete. New steels have also produced high-strength bolts, making rivets—and the chatter of the rivet gun—as passé as the bustle.

Other materials, too, continually alter the construction picture. Light metals, such as magnesium or aluminum, provide lighter concrete buckets or truck bodies, thereby giving bigger payloads. Plastics have already made a substantial impact on construction in everything from reusable waffle floor pans that form concrete floor slabs, to strong, woven translucent fabrics which can be draped over a heated work area during the coldest and windiest weather.

Concrete is also undergoing drastic changes. As the demand for faster-pace work cycles increases, concrete setting speeds become of paramount importance. Curing is the chemical transformation of the wet concrete mix into a hardened, strong mass. Concrete does not "dry" as it hardens; in fact it absorbs moisture which is made available during the curing period. Chemical acceleration additives aid in speeding the curing process; ultra-fast-setting cements, too, have been developed. Steam curing has rapidly become standard procedure in factories producing concrete products. Electrically accelerated curing—either externally through the forms or internally by imbedded heating coils—is gaining popularity, especially in colder regions.

The uses of concrete are evolving principally through the growing application of a single technique: pre-stressing. This process can be easily demonstrated. Take a row of books horizontally from a bookcase: try to pick up the entire row by grasping only the two end books. Impossible? Now try again, but this time tightly squeeze the row of books together, as when playing an accordion. The lift is now possible. That is an example of the principle of pre-stressing.

Concrete, then, is gradually changing its role from a weighty mass, necessarily bulky because of inefficient internal stress distribution, into a sleek, efficiently-utilized material capable of inspiring and meeting new design criteria, with load stresses strictly controlled by pre-stress action imposed by tightly stretched internal wires. One expert has predicted that before many years have passed, virtually every cubic yard of concrete cast will have some degree of pre-stressing in it. Already, self-stressing cements are on the market.

And now a polymer/concrete mixture, irradiated by cobalt 60, provides a material with four times the strength characteristics of concrete alone. Research developments along similar lines will undoubtedly produce numerous other "super-materials".

Heavy construction may be typified as an industry that traditionally makes the most efficient use of the natural materials at hand. Road-building is the best example of this. Yet often the natural materials along the selected route are inadequate for the anticipated loads. The gradation of gravels, for example, may not be quite right to guarantee the engineer stability under the continual pounding of heavy wheel loads. And the clay soils beneath may swell abnormally when wetted or thawed.

Nature's materials, then, need adapting. The designer may call on conditioning substances, such as cement, lime, calcium chloride or asphalts, to stabilize unruly soils. Or he may even prescribe treatment with a biologically active enzyme-type additive, which changes the characteristics of the soil. Continuing research in such directions—much of it slow, undramatic, methodical observation of actual field installations—will undoubtedly produce notable advances in the art of combining nature's products with man's.

New techniques in the field

For laymen, though, many of these important but subtle changes in design procedures and improvements in materials will be overshadowed—particularly in engineered construction—by technical and economic advances in field techniques, many of which admittedly are in response to design and material changes.

One of the most spectacular field techniques on the horizon is the use of atomic blasts to carve massive excavations in seconds. The United States Atomic Energy Commission is studying this in detail under its Plowshare programme. The economics of the method are attractive—preliminary estimates show rock or soil could be moved at about half the cost of conventional blasting, and much more quickly. But problems must be overcome: radiation dangers, the need to move people out of the blast area, international atomic treaties. The attractive benefits possible—for example, a new sea-level Panama Canal or possible diversion channels for Hydro-Quebec's James Bay power projects—may hasten the day when controlled nuclear energy does man's bidding for peaceful purposes.

Less spectacular advances in field techniques are constantly being made. Generally, because of the nature of construction, such advances must take place without benefit of laboratory testing or mock-ups. The testing ground is usually the construction site itself, and the whole profit or loss picture for the contractor often hangs on the success or failure of the new technique.

The success of the three-mile long conveyor belt used to move 57 million cubic yards of gravel to build British Columbia's W. A. C. Bennett Dam across the Peace River is an example of a contractor's ingenuity paying off. So, too, was the vast steel and plastic enclosure used to protect workers from −40°F. temperatures on Manitoba Hydro's Kettle Rapids job. Yet occasionally, the innovation becomes a money-loser instead of a saver. Hydro-Quebec tried, for example, to plug a diversion tunnel on its Outardes 4 dam project by tricky blasting of the ceiling. It looked good in theory but the tunnel failed to close and expensive remedial action was necessary. In another case, a mechanical tunneller made hesitant progress and eventually broke down, attempting to dig (without blasting) through several miles of solid rock for a new water supply line for Victoria, B.C.

Construction men learn from their own mistakes. Structural collapses and mechanical difficulties are scrutinized in detail to see what caused the failure. Mechanical tunnellers, for instance, will not suffer a permanent setback because of failures such as that at Victoria. With in-field data available, better design criteria are established for another try. Such tunnellers—nicknamed "moles"—are already, through this trial and error method, well along the road to providing contractors with a dependable, money-saving method for sub-surface excavation in both rock and softer soils. Similarly, such competing ideas as excavation with high-energy rays rather than rotating teeth, are being subjected to the rigorous testing ground of the construction sites.

Since any excavation in rock, clay, sand or organic swamp muck, is a key activity in heavy construction, improvements in excavating techniques will always be a principal aim of researchers and contractors alike. Modern dirt-moving equipment has already come far. In ancient times a single worker could move three cubic yards of dirt 500 feet in 12 hours. At such a pace the Egyptian pyramids and the Great Wall of China were built. One of today's modern scrapers, driven by one man, can handle more than 3,000 cubic yards in the same period. And who knows, before long, radio-controlled dirt-moving fleets, operated by one man sitting in a control tower, may be able to move a thousand times today's quantities.

Road-building has always been one sector of heavy construction where integration of various mechanical devices has been of paramount importance. Building a road is really a production-line operation, but instead of a product rolling out of a stationary factory, the "factory" of related vehicles moves along a predetermined route leaving the product—the completed roadway—in its wake. Modern equipment such as slip-form paving machines can place more than a mile of concrete pavement a day, two lanes wide, with a single machine performing the tasks formerly done by half a dozen rigs in sequence.

Asphalt paving manufacturers, too, already have prototype machines that pick up raw gravel in front, mix, extrude, and tamp an asphalt surface as they proceed, then add a dotted white line down the centre.

General Motors has projected what road-building rigs of the future may look like. Their demonic monster might very well have crawled out of the pages of a science-fiction novel. Grade preparation as well as pavement batching and placing would all be accomplished by a single pass of the machine as it moved along over virgin countryside. Machines will not be alone in this move towards integration. Designs by computer, we have seen, are already moving in that direction. And the construction companies themselves will, in the main, become integrated into larger entities. But even then, they will be only sub-sections of gigantic industrial complexes such as those found in Japan. The days of the individualistic construction entrepreneur are undoubtedly numbered. The "organization man" will ride the wave of the future.

General Motors of Canada, Ltd.

In this working model of the roadbuilding factory-on-wheels of the future, laser beams cut the forest while the automated machine excavates, analyzes and processes soils en route, and paves the roadway behind.

Just as construction itself will change, so, too, will the types of projects construction men will have to tackle. The hydro-electric dam's hold on power production appears to be waning as nuclear power comes into its own; the relative economics of the two generating methods swing in favour of the atomic-fired facilities. (In 1969 Ontario Hydro officials estimated that their Douglas Point nuclear generating plant, second largest in the world, would produce power at 3.9 mills/kwh, while their coal plants would produce at 4.7 mills/kwh. Water-powered plants can, depending on the site, produce power at lower or higher costs, but they do not have the advantage of being located wherever local demands are greatest). Even new types of power dams, designed to harness the tides such as those in the Bay of Fundy, have been evolved too late to compete with alternative power sources.

But in their role as respondents to demands for improvements to man's physical environment, construction men will follow, rather than lead, the demand trends. If better transportation is demanded, "slot-car" highways, or capsule trains and transports within pneumatic pipelines may become key segments of the construction market. Improved living facilities will be required on a world-wide basis. The response may be in the form of systems-built apartments, mile-high skyscrapers, domed cities or underwater hotels.

Although the construction industry, generally, has no real role in shaping specific demands, it does have an appreciable influence on whether or not those demands ever become satisfied.

For example, what if highway building were a hundred times more expensive than railway building, purely because of the construction problems encountered? The automobile, marvellous invention as it may be in itself, would of necessity have to stay parked at the factory (if, indeed, factories were built) because potential drivers simply would not be able to afford the taxes and tolls required to pay for the highways. Examples of this sort of economic interplay may be seen in the muskeg regions of Canada's sub-arctic, where economics of construction have long dictated that aircraft and ships (and more recently tracked vehicles and air cushion machines), not trucks, service the needs of scattered communities.

Projects of the future will include those in totally new environments— under water and in space. A thirsty world will demand construction of seawater conversion plants, and efficient irrigation systems to distribute the fresh water. Restraints on weather will come more by shaping the environment than by changing the weather itself. For example, frozen rivers and canals may be unlocked year-round by construction of integrated thermal-electric generating stations and river-warming plants. And domed cities will protect man from extremes of heat and cold.

The field for construction is as boundless as man's imagination. As surely as most of the needs and demands of past generations were met by constructors, so challenges of the future will in all probability be mastered.

BIBLIOGRAPHY

Highway for Today: Ontario's 401

Bank of Nova Scotia, "The Autopact In High Gear", *Monthly Review* (Toronto, June, 1969).

Canadian Imperial Bank of Commerce, "Urban Transportation in Canada", *Commercial Letter* (Toronto, January-February, 1969).

Mitchell, W. P., Assistant Controller, Financial Analysis, Ford Motor Company of Canada Ltd., "Canadian Automotive Industry".
(Address to Ontario Geography Teachers Association, Toronto, March 18, 1969).

Nowlan, D. and N., *The Bad Trip* (Toronto: New Press/House of Anansi, 1970).
(Criticises the Spadina Expressway, Toronto.)

Ontario Government, The Highway Improvement Act, 1960, Chapter 171, Section 38 (Toronto: Queen's Printer, 1967).

—— Information Section, Downsview.
(Various research documents, engineering studies, maps.)

Pleva, Professor E. G., former head, Department of Geography, University of Western Ontario, London.
(Excerpts from addresses and press conferences.)

Scott, Karl E., President, Ford Motor Company of Canada Ltd., "No Community Is An Island".
(Address to the Rotary Club of St. Thomas, Ont., March 3, 1966.)

United States Federal Highway Research Board, "Route Guidance", *HRB Record No. 265* (Washington: United States Government Printing Office, August, 1969).

—— "Economic Factors Influencing Engineering Decisions", *HRB Record No. 245* (Washington: United States Government Printing Office, March, 1969).

A City and Its Heart

Affleck, R. T., "Co-ordination and Organization of the Place Ville Marie Project", *Journal of the Royal Architectural Institute of Canada* (Toronto, February, 1963).

Barcelo, Michel, "Montreal Planned and Unplanned", *Architectural Design* (London, July, 1967).

Blake, Peter, "Downtown in 3-D", Architectural Forum (New York: Urban America Inc., September, 1966).

Brett, J. E., "Structural Design of Place Ville Marie", *Journal of the Royal Architectural Institute of Canada* (Toronto, February, 1963).

117

Cobb, Henry N., "Some Notes on the Design of Place Ville Marie", *Journal of the Royal Architectural Institute of Canada* (Toronto, February, 1963).

Montreal City Planning Department, *Centre Ville* (Montreal, various issues).

Montreal Gazette (Montreal, various issues).

Montreal Star (Montreal, various issues).

Philadelphia Redevelopment Authority, *Annual Report, 1968* (Philadelphia, Pa.).

Pinker, Donovan, "Planning", *Journal of the Royal Architectural Institute of Canada* (Toronto, February, 1963).

Ritter, Paul, *Man and Motor* (London: Pergamon-Collier-MacMillan, 1964). (Analyzes society's efforts to cope with the monster it has created in the automobile. Projects in many parts of the world, to provide separation of traffic forms and to make people more secure and comfortable in their environment, are described, as well as the historical growth of the concept. A staple on every city planner's bookshelf and available in most large Canadian libraries.)

Schoenauer, Norbert, "The New City Centre", *Architectural Design* (London, July, 1967).

Southwestern Legal Foundation, *Proceedings, Seventh Annual Meeting on Planning and Zoning* (New York: Matthew Bender & Co., Inc., 1968).

Zeckendorf, William, "Financing of Place Ville Marie", *Journal of the Royal Architectural Institute of Canada* (Toronto, February, 1963).

Industry on the Move

Federal Department of Labour, *The Labour Gazette* (Queen's Printer, Ottawa, January, 1970).

Heavy Construction News, Weekly news reports and feature articles on various aspects of techniques and ideas in present and future construction. Index available (from 1966) at Maclean-Hunter Ltd., Toronto.

McNevin, George H., *Heavy Construction News Management Study* (Toronto: Maclean-Hunter, 1966). (A series of six articles dealing with management in the construction industry. Reprinted in booklet form by the Canadian Institute of Quantity Surveyors, Scarborough, Ont.)

Royal Bank of Canada, *The Canadian Construction Industry* (Montreal, October, 1956). (Presented to the Royal Commission on Canada's Economic Prospects.)

INDEX